Countdown to Launch!

Rockets, Computers, and Coming of Age in the Cold War

Ralph L. Hollis, Jr.

ISBN 979-8-9903686-0-6 (pb)

ISBN 979-8-9903686-2-0 (hc)

ISBN 979-8-9903686-1-3 (epub)

Library of Congress Control Number: 2024905780

Butterfly Book Group, an imprint of Butterfly Haptics, LLC

In memory of
Beth Cawthra Hollis
She insisted my story must be told

Contents

Preface

This memoir chronicles the development of a young man's passion for rocketry and computers at an early age, later leading to his crucial work on software for a new intercontinental ballistic missile (ICBM).

It is a story about building rockets and launching them on the family farm near the Cold War target of Wichita, Kansas nearly two years before Sputnik. Curiosity, effort, and perseverance was needed to calculate the flight performance of those rockets using early digital computers. Marriage, young fatherhood, and abandonment forced aside the young man's goal of earning a Ph.D. in physics.

The story highlights the development of the Minuteman III ICBM, the first to have multiple independently targetable re-entry vehicles (MIRV), and the young man's role in developing the simulation software that would prove its capabilities. It is a story about a company, Autonetics, in Anaheim California, that grew to be the nation's largest military aerospace research and development complex, which now no longer exists. The story offers a peek into what it was like to work in one of the great aerospace companies of the 1960s.

For the careful reader, the memoir teaches for the first time the design of the Minuteman III missile in near-comprehensive detail, as well as the design of the digital simulator that verifies its operation. The general idea of the physics and mathematics involved is revealed without resorting to equations or program code. The more scientifically knowledgeable reader will hopefully excuse any simplifications. The book does not contain any classified information. Much of the information presented is publicly available, thanks to the Freedom of Information Act. Minuteman III remains the nation's sole land-based ICBM force. Final Epilog and Retrospective chapters bring the reader up to date on the current situation regarding these nuclear weapons. The work done over fifty years ago to deter war is still relevant today because of new threats of nuclear war.

My memoir is, of course, a true story and first-person eyewitness to history. I have represented the characters as I remember them. All are real people; there are no composite characters. In most cases it is a futile task to try to recount exact word-for-word conversations, but I have tried hard to retain their essence. Apologies if I have inadvertently missed the mark or slighted anyone. The main story takes place in the 15 years from 1955 to 1970. Writing the first draft started in January, 2019 through May, 2021.

I thank Francie Marais for her steadfast support, suggestions, and encouragement throughout the work. I thank Gloriana St. Clair, Patrick J. O'Connor, Ron Gallop, Stephen Hawkins, Greg Hopwood, Larry Hambly, and Jonathan Hollis for their careful reading of the draft manuscript and suggestions for improvement. Thanks to Oliver Kroemer for checking the readability of some of the more technical chapters. I am indebted to Julie Albright and William Boggess for their editing skills.

Telephone conversations with Audine Zuschek, Marc Roth, Phillip Roberts, Don Burton, Phil Rinard, Glen Reese, Georgy Santini, John Hlavac, Alan Bernstein, John Sweet, Lee Ann Romanizyn, Jon Freeman, Jean Anderson, Dave Dorius, Betty Jo Kraus, Bob Moffitt, Lucille Randolph, Keith White, and Al Sheue were extremely helpful in dredging up rusty memories from over fifty years ago. Thanks to E. William Guenther, Lt. Col., U.S.A.F. (ret.) for his helpful discussion of the B47 bomber.

Ralph L. Hollis, Jr.
Pittsburgh,
March, 2024

1 Countdown

D RIVING ON A WINDING DIRT ROAD with headlights off, Larry and I arrive a hundred feet from a locked gate to Vandenberg Air Force Base. I back my MGB into a grove of trees, then pull up to hide behind bushes. We don't dare speak.

Soon, a vehicle approaches the gate from the other side. A jeep with two guards stops ten feet from the gate, the beam from its headlights reaching out to us. After a minute, it turns around and heads back.

I feel my heart beating. "A close one," I whisper.

"Are we actually going to do this?" Larry asks.

"Let's go. I think we'll be all right."

I take a flashlight out of the glove compartment, and we step out of the car. We cautiously click the doors shut and accommodate to the darkness and warm pungency of the eucalyptus around us. It is quiet except for sounds of insects. Twenty feet to the right through the weeds, we reach a three-strand barbed wire fence. Larry reaches down and spreads the wires apart, and I slip through. Then, I do the same for him.

We head into the grassy field, hearing faint sounds of what must be cattle. We are here to witness the secret launch of a new kind of intercontinental ballistic missile with triple the destructive power of any of its predecessors. Larry is a key part of my team of programmers forming one small cog in the giant human machine developing this awesome weapon. Only top management had been invited to see the launch—from the safety of the blockhouse. My request was denied, but Larry and I had decided to see it anyway.

We make our way uphill through darkness for a quarter mile. Suddenly, I lose footing. We both stop. "I'm going to turn on the flashlight. This is insane. We could fall into a hole."

"Don't do it. Somebody will spot us."

I point the flashlight ahead of us and push the button for a few seconds, revealing a grassy incline for the next twenty steps, then blackness. As we slowly make our way up to the top of the rise, we see a ravine, looking to be thirty feet deep. "It's not bad. I'll keep the light pointed downward and we can make it down and up the other side." My stomach starts to tie itself in knots, thrilled to hopefully witness firsthand the results of our effort. On the other hand, I am terrified there might be some hidden flaw in the code causing everything to explode in a ball of fire.

It is the night of April 11, 1969. In the traumatic year before, the nation survived the assassination of Martin Luther King, Jr., followed soon thereafter by the assassination of Robert Kennedy. Riots tore through cities. People were gunned down in the streets in the greatest wave of social unrest in a century. The Hong Kong flu pandemic killed millions. The Vietnam War raged onward with massive casualties. A singular bright spot illuminated the world amongst the chaos: three men in an Apollo spacecraft were sent into orbit around the moon. Now, there are plans to land men on the moon before the end of this year.

Tonight, we are testing a weapon that could give the nation the means to prevent a nuclear war. But, in my darkest hours, I sometimes worry it could also destabilize the balance with the Soviet Union and help to trigger a nuclear war. Approaching midnight, the lights of Santa Maria appear to the north. We spot the fence in the starlight, the neat, straight line heading up the mountain, marking the base boundary. "We should stay at least ten feet away in case there are sensors that might detect us."

Now, gaining elevation, we can barely make out the western edge of the continent. Below us, only four-hundred yards away, is the unmistakable square pattern of red lights marking the outline of Launch Facility 02. We know the missile is poised to leap out of its silo. We wait, and presently we hear talking over a loudspeaker. I check my wristwatch—only fifteen minutes until launch. Someone calls out parameters and numbers related to the missile. Silence. I breathe deeply, sweating from the climb. My stomach tightens some more. Silence. Then, abruptly, the countdown begins.

I

WICHITA

2 Wrath of Vulcan

LAUNCH PREPARATIONS were underway in the cloudy morning of March 3, 1956. Steve Hawkins, Phil Roberts, Harold Wiebe, Gary Peyton, and I were ready to launch a homemade rocket from the middle of the alfalfa field on my parents' farm twenty-seven miles southeast of my home in Wichita. We were middle school students and founding members of the as yet unnamed science club meeting weekly in my parents' basement. I was president. The rocket was part of a project I planned to submit to the Future Scientists of America Foundation.

We started to set up our equipment at around 5:00 am. Now it was nearing 6:00 and the sun was coming up. We chose a location four hundred feet west of the barn to set up all the equipment. An electric excitement permeated the group as I began to coordinate the preparations. First, we set up and leveled the wooden launch platform at ground zero and started to get organized. A light breeze was blowing almost directly from the south.

"There's a big wooden panel in the barn we can use as a blast barrier. I'll need some help from at least two of you. Wiebe, how about you and Peyton?"

"Sure, chief!" I ignored Harold's sarcasm, thinking about how Goddard must have felt out in the New Mexico desert. Harold was taller than I, wearing horn-rimmed glasses and a blond crew cut. A true gadgeteer, he was interested in radios and all things electrical.

After a lot of effort dragging the seven-foot square board out of the barn and over the rough ground, we set it up thirty feet from the launch site. "Since I'm the photographer, I don't want to be behind the barrier with you and Peyton." Harold had brought his 35-mm camera and a surveying transit from his father's business.

"I just want everybody to be safe! I want you to stay back with Peyton and me," I said.

"Then how am I supposed to push the launch button?" Harold was the electronics expert in the group and had already built some Heathkit power supplies and voltmeters. "I'll be fine."

Harold set up his position lying on the ground behind the firing console a mere twenty-five feet southeast of the blast zone. Harold and I had designed and built the firing console out of a large wooden toolbox my father had made in college. It contained a battery, a Model T Ford spark coil I had purchased from Western Auto, a few small lights, a push-button firing switch—and most importantly, an old telephone socket. When installed in the socket, a telephone plug would complete the firing circuit.

"Hawkins, you can man a tracking station due east of the launch site. Set up with the altitude tracker a hundred feet away. Get some help with laying out a hundred-foot baseline." I had brought Dad's fifty-foot tape measure for this. Steve Hawkins was a tall and lanky guy with an intellectual bent. "There are some old wooden doors in the barn you can use for blast barriers." He would be using the altitude tracker with the tripod, setting his door in place vertically, supported with stakes I had grabbed from our backyard garden back in Wichita. "Roberts, you can man a second tracking station due south of the launch site." Phil Roberts would be lying on the ground behind his door which was set up horizontally. Steve and Phil were best friends. Phil lived a block away from my house—short brown hair, big ears, and one of the smartest guys I knew. Both were excited about rockets and science in general.

It was late morning when the sun peeked through the clouds, taking some of the chill off. The wind was starting to pick up a bit. I pulled the telephone plug out of the firing console, putting it in my pocket. At the launch site, Phil and I carefully removed a plaster cone from the center of the rocket combustion chamber, and a bit of powdered propellant fell out. The resulting conical void would provide a large surface area for burning. We set the rocket on its platform with its fuse hanging down a foot into a small mound of powder. At last, after a painstaking building process, my rocket, White Vulcan, stood proudly on the launch stand. The spark coil and battery were set up next to the rocket behind a plywood barrier, and the ignition wiring was unrolled and connected to Harold's console.

"Here's the plug." Harold took it from me and plugged it into the console. The system was now "hot."

"Prepare to launch!" I yelled out the names of the launch crew one by one

6

and each man replied, in turn, that all was ready. It was time for the countdown. Gary called out "10, 9, 8, 7, 6 ..." At zero, Harold pushed the launch button. Nothing happened. After several more attempts, we knew we'd have to ignite the fuse by hand.

"I'll do it," said Gary, taking the telephone plug out. Gary Peyton, short brown hair with glasses, lived in an old Wichita farmhouse a block away from me. His parents were divorced; he was being raised by his mother. Gary had a love for chemistry. Matches in hand, he walked the thirty feet to the rocket, ignited the fuse, and ran like hell back toward me and the barrier.

Suddenly, the bottom half of the launching stand was engulfed in flames. The fuse had disintegrated, and a deep steady roar filled the air. A foot-long semi-transparent blue-white flame of exhaust gases neatly fired out of the tail of the rocket. Amid huge clouds of billowing smoke, the rocket lifted six inches off the stand, tilted sideways, began to lose altitude, and partly supported by its pillar of flame, hit the ground five feet away. It was momentarily obscured by smoke, but I could tell it had lost one fin. The rocket then shot across the cloddy ground a few feet northwest, then southeast for fifteen feet, then suddenly turned south, straight for Gary, who had just made it back to the barrier. It stopped five feet in front of the barrier. Everyone ran up to the rocket, and we put out a small flame coming out of the combustion chamber. As we stood there, I tried to hide the disappointment from my face.

Dismayed, I concluded that my White Vulcan failed to reach appreciable altitude because the thrust of the motor was not sufficiently greater than the weight of the rocket. Its low initial velocity made it a prey to the winds. I had said shortly before the launch, "It will fly." But it did not.

I wrote in my project notebook: *"Perhaps the next rocket will fly. And if not the next, perhaps the one after...but someday a rocket will fly."* Nevertheless, I finished the documentation of my failed project and mailed it to the Future Scientists of America Foundation in Washington, DC.

* * *

Many events led up to the failed launch of the White Vulcan rocket. On July 29, 1955, President Eisenhower had announced plans for the building and launching of the world's first artificial satellite. Earlier, S. Fred Singer, then of the University of Maryland, had designed the MOUSE (Minimum Orbital Unmanned Satellite of the Earth) as a refinement of a 1951 proposal in the Journal of the British Interplanetary Society for the smallest possible device that could perform useful scientific research in space. My imagination was captivated when I saw a picture of the MOUSE in a magazine, in what may have been as early as 1954. It became a hot topic of discussion in our science club.

The previous April, I had read some popular magazine articles by famed meteorite hunter and astronomer Dr. Lincoln La Paz, who happened to have been born in Wichita. He and astronomer Clyde Tombaugh, the discoverer of Pluto, postulated tiny "moonlets" would be discovered circling the Earth at high speeds. Further, they said these moonlets would provide excellent bases for space stations. As a thirteen-year-old kid, I strongly disagreed with this idea, so I wrote a two-and-a-half-page letter listing seven separate arguments opposing the writings of these famous men. Instead, I touted Dr. Singer's approach to artificial satellites. Alas, I never received a reply.

As I began the 9th grade at Charles Curtis Intermediate School, my science teacher, Joe Foraker, suggested I enter the competition sponsored by the Future Scientists of America Foundation. It sounded to me like a good idea. For my entry, I decided to make a detailed full-size model of Singer's MOUSE.

I started work building a model of the MOUSE in September 1955. Its main structure was a one-gallon fruit can I carefully cut away to "reveal" the internal components, which were modeled in balsa wood with pieces of wire and other scrap electronics intended to represent cosmic ray and gamma ray counters, ultra-violet and solar X-ray detectors, a magnetometer for measuring the Earth's field, and micro-meteoroid impact detector—all faithful to the picture of Singer's proposed satellite.

Once my MOUSE project was well underway, attention turned to building my own small rocket. I realized it would take a huge rocket to launch the real MOUSE into an orbit around the Earth. No rocket had ever achieved such a feat. My rocket would serve only as a modest illustration of rocket propulsion principles. I wanted to take a safe and scientific approach. As a first step, I took the bus to the Wichita City Library where I found a book called *Rockets and Jets*, published in 1945. Amid diagrams of Chinese rockets, I discovered the best fuel mixture, so-called "black powder," consisted of 75 parts of potassium nitrate, 15 parts of charcoal, and 10 parts of sulfur. But I also read this mixture could be too rich. It could burn too fast, producing high internal pressure resulting in an explosion. So I thought I might try a mixture with less potassium nitrate. My first task was getting my hands on these ingredients.

Gary, the chemist of our group, would purchase the chemicals. He warned me I would have to use caution in grinding the potassium nitrate powder to be sure that no burnable impurities were in the mortar and pestle. I got the sulfur at Maule Drugs, where it was sold as "Flowers of Sulfur." Gary soon procured everything else.

I decided the hull of the rocket could be made from a cardboard mailing tube I got from Dad. I coated the inside of the tube with a solution of sodium silicate to make it more heat resistant. To test its fire-resistance qualities I made three small samples of laminated wood, tin, and cardboard, coating each of them with the sodium silicate solution and letting them dry thoroughly.

At our November 16 science club meeting, I exposed each of the test samples to a candle flame. To my disappointment—and the surprise and delight of the other members—the cardboard sample fizzed and cracked and began to char in exactly seven seconds. The other samples fared a bit better. From these experiments, it became readily apparent that my thin cardboard tube would neither be able to hold the intense heat nor the high combustion chamber pressure. The rocket would probably explode shortly after igniting it. Four days later, Phil secured a one-eighth inch thick mailing tube that hopefully would be sufficient to withstand the intense heat and explosive reaction.

I worked on the MOUSE and my rocket steadily into January 1956, fashioning the fins—modeled after the fins on the German V2 rocket—and a carefully carved nose cone from balsa wood purchased from Hillside Hobby Shop. I painted my rocket white with a bright red band around its center section, christening it "White Vulcan."

My battery-powered Model T spark coil had a vibrator sounding like a nest of angry bees as it generated an impressive spark from ten thousand volts of electricity. Its radiated energy could cause nearby fluorescent tubes to turn on and flicker. It would be a perfect ignition source for the rocket.

During cold and snowy January, I performed a series of tests out on the backyard patio to find a good propellant mixture. The mixtures were ignited at a distance with the spark coil. I needed to find a good mixture soon because the submission deadline for my project entry was coming up in early March. After the twenty-second test, the powder flared up tremendously and disintegrated with great heat, starting several small fires in the area. I figured if this fuel were properly installed in a rocket it would be quite sufficient. The final mixture was 62 parts potassium nitrate, 25 parts charcoal, and 13 parts sulfur.

I began to turn my attention to the matter of tracking the rocket's flight. I made a simple "Altitude Tracker" from plywood cut out with my small Craftsman jigsaw (the best Christmas present I ever got). A hinged pointer could be tilted to follow the rocket visually and locked in place with a wing nut when the rocket reached its peak altitude. The pointer moved alongside a protractor with a scale going from zero to ninety degrees. The base of the apparatus could be swiveled horizontally. It could be mounted and leveled on Dad's tripod, part of his surveying equipment. Given the distance of the tracker from the launch site and the recorded angle, I figured I could make a scale drawing which could tell me the achieved height of the rocket's flight. (I could have figured this out using trigonometry, but I was still a few years away from learning that in school.)

Phil helped with loading the solid fuel into the rocket at 9:00 the night before the firing. I was concerned we might accidentally burn down the house, so we did this work on the floor of the garage. Following pictures from the *Rockets and Jets* book, I had previously made a thin cone out of paper with a soda straw down the center. I poured a thick Plaster of Paris mixture into the annulus between the straw and the cone and let it set up the night before. We made a fuse out of soft cotton twine dipped into a potassium nitrate solution. Then, we rubbed the string through the black powder fuel mixture. After the fuse had dried, we put it through the straw and inserted the cone with its straw into the rocket tube.

Then, it was time to load the rocket fuel mixture into the rocket. We moistened the mixture in a coffee can and spooned it into the top of the rocket a little at a time, pounding it down with a wooden dowel, expecting it might blow up at any second. We embedded the fuse in the propellant at the top of the cone. When the level of propellant reached the top of the rocket body, we re-installed the nose cone with a tin heat deflector plate, small nails, and glue. Finally, we packed the loaded rocket, cushioned with rags, into an old wooden box along with the other equipment and left it in the garage until morning. As I fell asleep, I wondered if I had created a lethal bomb.

* * *

Now, after the spectacular failure of White Vulcan, it was June and I could hardly wait for the grasp of school to end, releasing me into the summer. It was the end of Charles Curtis Intermediate School for me. Next year I would be a 10th grader at East High.

At the honors assembly, a boring series of football players, basketball players, and band members marched across the stage to collect their certificates. Surprisingly, Mr. Foraker called me up to the stage to announce that I had won first prize in the six-state region for my science project! I received a $25 government bond, a certificate, lapel pin, and a plaque. Apparently, the judges had valued the analysis of failure as much as if the rocket had flown into the sky. A day later, I was surprised by an article, *"Model of Man-made Satellite Wins Future Scientist Awards,"* and picture in the Wichita Eagle, plus a smaller article in the Wichita Beacon. I was thrilled at becoming a little bit famous.

* * *

In summer, my mind continued to be occupied by dreams of rockets. White Vulcan II was taking shape in the basement workshop. The rocket's hull had the same thickness tube, but also had a molded clay nozzle at the end, baked in Mom's oven.

I had read about so-called "de Laval" nozzles. The nozzle was developed (independently) by German engineer and inventor Ernst Körting in 1878 and Swedish inventor Gustaf de Laval, in 1888, for use on a steam turbine. The principle was first used in a rocket engine by Robert Goddard. As hot turbulent gases escape the combustion chamber they speed up and reach the speed of sound near the nozzle's "throat," and then speed up further as they escape into the air. The momentum of the escaping exhaust gases imparts an equal and opposite momentum to the rocket, which propels it skyward. This would be a vast improvement over the first White Vulcan, which did not have a nozzle.

Another improvement was in the launch stand. Now, four wooden rails surrounded the rocket, stabilized by guywires. The rails were about four feet in length and would guide the rocket during its initial flight. The idea was that when the rocket left the top of the rails it would be going fast enough that its fins would provide stability against random wind gusts.

It was early July when my crew was joined at the farm by my brother Dave and Wendell Allen. Dave, two years younger than I, was a loyal helper. Wendell was not a member of our science club but had asked to join us for the launch. He seemed intelligent but not as focused as the rest of us. It was 5:00 in the afternoon before all the equipment was set up and ready. A gentle south wind blew over the area as the final countdown was made.

This time, the spark coil ignition system worked perfectly. As soon as I pressed the button, the rocket ignited. A bright splash of orange flame and three small staccato explosions ensued. I could tell at that moment my summer's work would never leave the ground. As the seconds ticked on, the launch stand became a roaring inferno—which continued to burn for another five minutes. Nothing was salvaged except a portion of the nose. Once again, it became evident that something very basic was wrong with the way I was building rockets.

* * *

For the rest of the summer, I labored at building White Vulcan III. Instead of a fuse and cone used in the previous Vulcans, the black powder was loosely packed in, and a perforated cardboard retainer was installed a short distance forward of the convergent section of the exhaust nozzle. According to my notebook, the nozzle was made of "Vulcanite," a non-metallic mix, whose composition is lost to history. This time, the ignition system was a kerosene-soaked wick to be ignited by the spark coil. The previous wooden guidance rails destroyed by fire were replaced by four steel rods.

It was 3:00 in the afternoon of October 5, 1957. A moderate east-southeast wind prevailed. As my third Vulcan rocket sat poised on the launch stand, the final countdown began. The button was pushed, followed by a delay of at least ten seconds before the rocket ignited. A roaring ball of fire covered the launch stand. In three seconds, the rocket was doomed. Total burning time: five seconds; altitude: zero. Vulcan, the Roman god of fire, had struck again.

In the evening, back in Wichita and feeling discouraged at the failure of White Vulcan III, I got a phone call from a breathless Wendell Allen. "Hey Ralph, have you seen the news? The Soviet Union has launched an artificial satellite! They call it 'Sputnik'." I didn't know it then, but the world had changed forever.

3 Sputnik

STEVE HAWKINS AND I headed to the Moonwatch site on Wendell Allen's parents' farm north of Central Avenue, several miles east of Wichita. The day before, on Friday, October 4, the Russians had launched the world's first artificial satellite. My brain was on overload with excitement as I pulled into a parking area.

As part of the 1957-58 International Geophysical Year (IGY), the Navy would launch a space satellite using its Vanguard rocket. I had eagerly read newspaper and magazine accounts of the project, cutting out articles and pasting them in my scrapbook. Operation Moonwatch was the brainchild of famous astronomer Fred Whipple. Teams of amateur observers would view the Vanguard satellite as it passed over their locations, using small telescopes. They would mark the exact time the satellite would cross the Zenith, and report this by telephone back to Moonwatch headquarters. By aggregating these data from all over the world, the satellite's orbit could be calculated. Now, with Sputnik, the Russians had beaten us to the heavens, and Operation Moonwatch was tracking a foreign satellite.

As my eyes got used to the darkness I could make out a dozen men in uniform from nearby McConnell Air Force Base who were engaged in a soft buzz of conversation. Some people were sitting on folding chairs arranged in a row in front of small tables. Small telescopes with mirrors on each table enabled the observation of wide areas of the sky. In the distance, barely audible, was the clanking and sucking of oil well pumps taking their sips from the earth.

Soon, I spotted Wendell. I asked him where people got the small telescopes; he told me you could buy them from Edmund Scientific, in New Jersey. Steve and I hung around until after midnight. No one there spotted Sputnik that night.

The next day's Eagle and Beacon newspapers were awash with reports of the new Russian achievement. Some of the headlines and stories in the Wichita Eagle, dateline October 5, 1957: "Russia Claims First Earth Satellite Launching," with a drawing showing the predicted orbit passing over the United States; "U.S. Radio Hears 'Beep', Satellite's Signal Picked Up by RCA"; "Red Satellite Believed Seen Over Ohio"; "'Beeps' from Red Satellite Heard by Listeners Here," reported by local ham radio operators. I cut out these historic articles, carefully pasting them in my scrapbook.

On Monday, I watched a newsreel on TV narrated by Ed Herlihy. It was short, consisting of animated drawings with the headline "Reds Launch First Space Satellite."

"Today a new moon is in the sky. A 23 inch metal sphere placed in orbit by a Russian rocket. Here, an artist's conception of how the feat was accomplished. A three-stage rocket. Number one, the booster in the class of an intercontinental missile, its weight estimated at fifty tons. A smaller second stage took over at five-thousand miles an hour and carried on to the highest point reached. Five-hundred miles up the artificial moon is boosted to a speed counterbalancing the pull of gravity and released. You are hearing the actual signals transmitted by the earth-circling satellite, one of the great scientific feats of the age."

On October 9, in a prepared statement, President Eisenhower announced the United States would start firing Earth satellites into outer space in December. "As to the Soviet satellite, we congratulate Soviet scientists upon putting a satellite into orbit." In a news conference, the president expressed that the Russian satellite was no threat to U.S. security. He also questioned Sputnik's reported weight of 180 lbs.

On October 11, Air Force spotters stationed in Alaska photographed the streak of light from the third stage of the Russian rocket which remained in orbit closely following the new moon. Under the headline "Mechanical Brain Fixes Orbit of Moon's Rocket," was a story, *"An IBM mechanical brain at MIT, in Cambridge, Massachusetts, working nightlong, determined the orbit on the basis of photographs and observations of the rocket which trails the world-girdling satellite."*

The Soviets did not release a photograph of Sputnik until five days after the launch. Until this time, its appearance was a mystery. Sputnik's weight of over

180 lbs., compared with the Vanguard's proposed satellite weight of 21.5 lbs., led many to doubt the Soviet claim. Our biggest operational rocket, the "Redstone," weighed 61,000 lbs. and produced approximately 78,000 lbs. of thrust. U.S. officials calculated the Soviet rocket that launched Sputnik would need to produce at least 200,000 lbs. of thrust to launch the satellite into orbit. It was later learned, but not disclosed to the public, that the R7 rocket that launched Sputnik into orbit weighed around 560,000 lbs. and produced a thrust of almost 800,000 lbs.

I read a short opinion article printed in the Eagle: "Effect on Man by Earth Satellite Hard to Foresee" by J. M. Roberts of the Associated Press.

"Those ghostly little beeps which have been coming out of the heavens since Friday offer man more food for thought than anything since unlocking the secrets of the atom. It is impossible even to list at one sitting all the facets of the advent of the man-made satellite. Soldiers, historians, scientists, diplomats, philosophers and just plain men will be coming up with new angles on the subject for a long, long, time. There is a sense of human accomplishment in this break through natural barriers which have intrigued the imagination since the creation."

Not generally recognized at this point by ordinary people was that the power of the atom and the power of the rocket would soon be united.

4 The Red Balloon

NATURE OR NURTURE—which dictates our destiny? Plato suggested that important traits are inborn, occurring naturally regardless of environmental factors. On the other hand, John Locke advocated a *tabula rasa* point of view, concluding that we are born a blank slate and that our knowledge and personalities are determined by our experience. In my case, I believe that a few early events helped to influence the eventual man I was to become.

I was born in Hutchinson, Kansas, ten weeks before the Japanese attack on Pearl Harbor. When I was a year old, we moved to Santa Fe, New Mexico, living in an adobe house a block from the historic Plaza. Dad worked as an architect in a second story office overlooking the Plaza where he made drawings of wooden barracks and bridges for a wartime project to be built atop Los Alamos mesa some 40 miles northwest of Santa Fe. After a year or so in Santa Fe we moved to Fort Worth, Texas, and then to Salina, Kansas, where my brother David was born a few days before I turned two.

My first memories are from around age three when our young family moved into a clapboard duplex apartment in Wichita, a block from the Hilltop Manor development built to house wartime workers. Wichita surrounds the confluence of the Arkansas and Little Arkansas rivers. Situated between the rolling Flint Hills to the east and the flat prairie land to the west, it was home to the Quivira people who spoke the Wichita language. After the Civil War, it was a trading post along the Chisholm Trail. Cattle were driven up from Texas through Wichita to the railhead in Abilene. The famous lawman Wyatt Earp moved to Wichita in 1874.

Of course, I didn't know any of this as a small boy. Even at that age, I was aware that Wichita was surrounded by vast fields of wheat for as far as one could see. The flat land lay under a huge, wonderful sky. Lying barefoot on my back in the cool grass, I spent hours gazing at fluffy white clouds turning them into lions and birds. I also understood that Wichita was all about airplanes. Lots of airplanes!

The continual roar of bomber engines and machine gun fire, coming from the Boeing plant less than four miles away, went on day and night. One afternoon, while playing in the front yard, the roar got extremely loud. Looking up, I was astounded when a giant B-29 bomber appeared low overhead in a left bank with its right inboard engine on fire and trailing smoke. I followed it as it disappeared in the distance. In those years, a massive human force was at work making B-29s. I later learned that my young cousin Wanda, from a farm near Herington, Kansas, had been a "Rosie the Riveter" at the Boeing Airplane Company.

My world was small. The apartment was graced by a tiny living room with a pot-bellied coal stove, a tiny kitchen, bedroom, and bathroom. We brought in buckets of coal to stave off the chill — of what Mother called "Old Man Winter." Two mean older boys lived in the other half of the duplex; a little girl I admired played in the front yard of the adjacent building. A nice man two doors down had only one eye. He had been operating a lathe in an aircraft factory when a piece of metal flew off and took out his eye.

Across the dirt street was a forbidding field of weeds and a big cottonwood tree. A narrow sidewalk turned left, sloping slightly downward from our front stoop, turned left again to follow the side of the house and ended at the gravel alleyway. The coal bin faced the alley. I could walk and balance on its top edge but stopped trying after I fell in, gashing my forehead on a sharp chunk of coal. The alley was a frequently explored treasure trove of detritus. I rescued a pretty blue slightly cracked dish and wrapped it with some tissue paper for a Mother's Day gift. In summertime, we had a vegetable garden in the backyard where I picked raw peas that popped in my mouth with delightful sweetness.

We didn't have much money, but we were never poor. Father worked for a while at Boeing, and to my delight brought home paper airplanes from work. I was fascinated by how they glided through the air, imagining I was aboard. He gave me a small magnifying glass that allowed me to scrutinize bees and flowers up close. I credit this to the genesis of my science career. I saw little of my father during those years. Part of the time he was putting in overtime hours at Boeing extending beyond my bedtime. But for a big stretch he was in faraway Washington State in a place called Hanford, on the Columbia River. Mother endured stoically, repeatedly joking that Father wouldn't recognize us when he got back.

On the front stoop one warm summer evening, Mother sat on my left with her legs folded up as we watched lightning bugs dance and fly. A line of storms flickered orange in the west. I asked why we couldn't hear any thunder and she replied it was only heat lightning. I asked if Father was under it, afraid that he would be hurt, and she told me not to worry; he was behind it—much, much farther away.

Next to the coal bin we had a hutch that housed two large rabbits. In due time the rabbits had six baby rabbits I adored and played with. We took one into the hospital—which was against the rules—to comfort my seriously ill older cousin. She was delighted. As the rabbits got bigger, I noticed they were disappearing one by one. Concerned something or someone was stealing the rabbits, I was told one evening at dinner the horrifying truth about the missing rabbits.

To my delight, after a cold snowy morning, Mother built a huge snowman in the front yard with coal eyes and a carrot nose. The boys from next door pushed it over and stomped on it. Another time, Mother tied pieces of string between my toys to make a long train. I was fascinated with playing with my new toy until the boys came over and maliciously snapped all the strings.

One summer day, I was staring out the living room window when I saw a big red balloon floating in the soft breeze, moving down the block. I had to have that beautiful balloon! I snuck out of the house and ran after it along the edge of the empty field. I was forbidden to leave the yard, but I didn't have time to tell Mother. It was only me and the balloon. I made it to the end of the block and across the street where the balloon had at last come to rest against a fence. I was so proud when I carried it back to the house and presented it to my worried mother. She took it into her hands, and it popped! I immediately burst into tears and started sobbing "Oh Mom—Oh Mom—your fingernails are too sharp! You broke my balloon!" This could be the origin of persistence followed by disappointment emerging as a theme in my life.

Several months before my fourth birthday, a B-29 bomber halfway around the world dropped an atomic bomb on the city of Hiroshima. Three days later, another atomic bomb was dropped on Nagasaki. The war in the Pacific was over; I remember hearing about it from my parents.

The atomic bomb had been designed at Los Alamos and tested at a place called Trinity in the desert near Alamogordo, New Mexico. The Hanford B Reactor produced the Plutonium used at Trinity and in the bomb detonated over the city of Nagasaki. Because of my father's work for the projects at these locations, there is a thin thread connecting these events with my own story.

It was a full moon, and I was out back by the coal bin. The mean boys from next door grabbed me by the collar and held me, forcing me to look at the moon's reflection in the fender of a neighbor's car. One boy yelled, "See that! It's an atomic bomb and it's coming straight for you. Will you die of fright or die from the explosion?" They scared the wits out of me. I jerked myself free and ran into the house. Up to this time I had not considered the notion of dying at all. Some twenty years later the question of whether one would die of fright or die of the blast would reappear in my life as something the news would call Mutual Assured Destruction.

5 Duck and Cover

GRADE SCHOOL was a happy time for me. Thomas Jefferson, adjacent to Oliver Street, was at the eastern city limit of Wichita. We fed grass to the horses through the fence bordering the small pasture east of the school building. In those days, morning and afternoon recesses were the best parts of the day.

In first grade, we got to eat graham crackers and take naps on our rugs on the floor. I was trying to see if I could cut a cracker in two by sawing through it with another cracker. I could cut a groove in a stationary cracker held in my left hand while sawing back and forth with another cracker held in my right hand, which worked well. Then, curious, I wondered what would happen if I sawed both crackers back and forth at the same time. Perhaps this experiment engendered my first notion of axes and coordinate frames that would occupy my life decades later. Before I could complete the experiment, the teacher came up behind me, suddenly slapping my hand hard enough to sting—sending broken crackers all over my table. Other kids who misbehaved felt the wooden paddle she had in a back room.

Bob Bullock was a friend in second grade. His father, who ran an electrical engineering service, would add electrical wiring to my father's architectural drawings. Imitating them, I would sketch plan views of fantastical buildings and pass them to Bob, who would return them a day later with all the wiring runs penciled in.

Later, Bob and I would raise pigeons in our back yards and build small labs in our basements. Inspired by seeing *Young Tom Edison* starring Mickey Rooney at the Tower Theater, my lab had glass bottles, salvaged from the alley three blocks away in back of our former duplex apartment. I had chemicals like salt and sugar, and liquids with food coloring on small shelves I had built into the corner of a basement closet.

Bob and I joined forces, forming B&H (Bullock and Hollis) Labs. We pursued many interests: building crystal radios, wiring up high voltage supplies to vacuum tubes, trying to build a jet engine out of a can, and working with magnets. Our jet engine lit off with a roar and twelve-inch exhaust flame, but then it exploded—covering my forearm with burning alcohol. The fire was soon out, and I suffered no lasting damage save for a few burnt hairs. These kinds of activities persisted for several years. I spent so many afternoons and evenings at his better-equipped lab that my father asked me if we were planning to get married.

In the early 1950s, after my grandfather passed away, Father used part of the estate settlement to purchase a 160-acre farm southeast of Wichita, near the small town of Douglass. Raised on a farm near Salina, he was looking to recreate his childhood days. Even though Dad was an architect by profession, he loved it best when he had a bit of dirt and grit between his teeth. My brother Dave and I were put to work on weekends plowing and harrowing the fields with our John Deere tractor. Father bought a horse, named Debbie, for my sister Karen, six years younger than me. Baby brother Brian came along in 1952.

Ronnie Gallop was the next-door neighbor kid, two years younger than I, the same age as my brother Dave. Without a sibling, Ronnie was continually over at our house, getting into all sorts of mischief with Dave and me. Short, with red hair and freckles, he was a curiosity. His father, Dwight, was a hardworking manager at a manufacturing company in Wichita. His build was a bit on the stocky side, and I don't believe he ever said much to me. Dwight had a modest workshop in the basement and was an innovative do-it-yourselfer. He had built a whole-house air conditioner from scratch and painted his roof silver to reflect the intense Kansas summer sun. One time, he suffered a serious hand injury from a work accident. He emphasized to us boys to be careful using tools. Ronnie's mother, Opal, was a housewife who didn't seem to mind us running in or out of her house at most any time. She had a record playing the song *"Detour, There's a Muddy Road Ahead,"* night and day.

Ronnie was a bit reckless. On a cold and wet morning, he flagged down the milk truck to buy some bottles of milk. Running back on the slick driveway, he slipped and fell face down on the glass bottles which broke into several large, sharp pieces digging deeply into his chest. Another time, fooling around with his father's gun, he shot a hole in his bedroom ceiling. After that incident, he made his own small gun out of an aluminum tube, powered by the explosion of a firecracker, with a short aluminum dowel for a bullet. He called me over to witness the test. We fired it off in his garage, where it shot completely through the door and likely hit the house across the street.

Ronnie came with us on our many trips to the Hollis farm, boating and fishing with us on the Walnut River. In turn, we made trips to his parents' cabin, where we tubed on the shallow, lazy, Ninnescah River. Down at our farm, he discovered saltwater conducts electricity when he peed on an electric fence. *Yeeow!*

He was perpetually into various adventures. He took guitar lessons, aspiring to become a star. He became a Ham Radio Operator, using his Hallicrafters equipment and 10-meter whip antenna to communicate with the world. We participated in "transmitter hunts," where a club member in the local Ham Radio club would hide a radio transmitter at an undisclosed location somewhere within the city limits, while the other members careened through the streets holding small loop antennas out of their car windows to detect lines of maximum signal strength. Since the antennas could not discriminate between pointing toward or away from the transmitter, this required making at least two readings from different locations to determine where the lines intersected. I usually rode in the back seat and did the plotting on a map. It wasn't easy to find the transmitter, since it was only set to transmit periodically.

* * *

Miss Wood, my third-grade teacher, was nuts about birds. She spent many hours in class playing records of bird calls, testing us on recognition. She had flash cards with beautiful Audubon bird drawings, drilling us until we knew each bird by heart. She took us to Riverside Park where we saw a tree filled with goldfinches. Her bird lessons took so much time she was unable to teach us multiplication, a required third grade skill. She got into trouble for this, but I am forever grateful to her, because it taught me it is okay to pick a single subject to study in depth.

In fourth grade, I met Phil Roberts who became a life-long friend. One afternoon, all the classes met out in the schoolyard to watch a traveling cowboy performer use the crack of a bull whip to extinguish a cigarette held in his pretty wife's lips. This is the kind of excitement we had in those days.

Sometime in fourth or fifth grade, we toured the salt mines in Hutchinson, Kansas. Six hundred feet below the earth, we took a small train through wide boulevards cut out of gleaming white salt. This experience, as well as a tour of the Wonder Bread factory, left lasting impressions.

In 1952, I was in sixth grade. Phil and I had Miss Warner, who liked science a lot. Each student had to do a science project, showing it to the others in class. I made a baking-soda-and-vinegar fire extinguisher I shot off into a waste basket to the approval of my classmates.

Sue Ann Bowling, a fellow sixth grade classmate, a tall awkward girl, brought in several Collier's magazines, with articles by Wernher von Braun and Willy Ley, illustrated by Chesley Bonestell. I was immediately struck by the visions of rockets to the Moon and Mars. I borrowed the magazines, studying the detailed plans and diagrams. Later, I made my own paintings of spaceships and alien worlds.

I saved up my twenty-five-cents-a-week allowance to buy a small microscope to study plants and insects. I got a chemistry set, at my eleventh birthday party, from Ken Kawaichi—son of Japanese Americans, who were likely interned in a concentration camp during the war. I defended him when he was attacked by a bully and called a "Jap."

Out of all possible future careers, I decided right there and then, in the sixth grade, that physics was the best. I desperately wanted to know how things work. I learned that physicists study nature in all its details. That was for me.

* * *

The world situation was looking dire in the mid-fifties and early sixties. When I was eight years old, President Truman announced on the radio, "We have evidence that within recent weeks an atomic explosion occurred in the U.S.S.R." At this time, the U.S. had tested five atomic bombs since Nagasaki. The ominous Soviet development was a new milestone in the developing Cold War.

Atmospheric tests of atomic bombs proceeded rapidly at the Nevada test site, with a total of twenty-four fission bombs tested by mid-1952. A considerable quantity of fallout from these tests drifted eastward on prevailing winds, blanketing many states, including Kansas, with a witch's brew of radioactive isotopes. Hundreds more tests were conducted until a pause in late 1958 that lasted for several years.

We had a special classroom movie called *"Duck and Cover,"* showing us what to do in case we were bombed. The film used a cute turtle to illustrate that you should immediately duck under your desk when you saw the flash of the atomic explosion, then cover your head and neck to protect yourself from flying glass. I was scared enough to take the advice seriously, and its message could have saved lives. In many instances, people survived the attacks on Nagasaki and Hiroshima at basically ground zero. (Later in my life, I knew a Japanese scientist whose grandparents survived the Hiroshima bombing and fled to Nagasaki in time for the second atomic bomb, which they also survived.) In 1952, atomic war did not necessarily mean instant or even prolonged death.

By late 1952, a hydrogen thermonuclear device of a type proposed by physicists Edward Teller and Stanislaw Ulam was detonated on Eniwetok Atoll in what is now the Marshall Islands. The device, called Ivy Mike, designed by physicist Richard Garwin, yielded over 10 megatons of TNT explosive power, vaporizing the atoll and creating a crater over a mile in diameter. (Much later in life, I had the fortune of meeting Ulam in Colorado and Garwin at IBM in New York.)

Ivy Mike showed the possibility that nuclear weapons could be built with a thousand times more destructive power than the bombs that obliterated Hiroshima and Nagasaki. Less than a year later, the Soviets detonated their first hydrogen device, with a yield of approximately 400 kilotons. In 1954, the U.S. Castle Bravo test in Bikini Atoll yielded over 15 megatons. By 1955, the Soviet Union had developed a deliverable hydrogen bomb in the low megaton range. "Duck and cover" was no longer an option.

On TV, President Eisenhower, striving for calmness in his voice, talking about the need for robust civil defense said, *"This great chore you have here is to give people the facts. Show them what they can do. Get the federal leadership, get the participation of the states and the municipalities without terrifying people. I have one great belief. Nobody in war or anywhere else ever made a good decision if he was frightened to death. You have to look facts in the face but you have to have the stamina to do it without, uh, just going hysterical."* My friends and I had other plans.

On my walk to and from Charles Curtis middle school, about a mile from home and three miles from the Boeing and McConnell Air Force Base presumed targets, I would make mental notes about where I would run for shelter from the inevitable nuclear blast. In most cases the best option was to squeeze into the nearest storm drain. I figured at least I would be below ground level. After testing this option a few times, Phil Roberts, Steve Hawkins, and I began to explore and map out a whole network of reinforced concrete underground sewer mains in the southeast part of Wichita.

It was an interesting time for me in school. I liked science and drawing pictures of airplanes. Math class disagreed with me. Math meant using simple

25

arithmetic to solve word problems. Bored with numbers, I had been looking forward to the ninth grade, where I could use letters like X, A, B, and C. But, because of poor math grades, to my disappointment I was not allowed to take Algebra. Instead, I was deemed unlikely to be college material, and was transitioned to a vocational track that included classes in shop and mechanical drawing. Actually, these classes suited me fine. In another of these classes, Vocational Studies, I wrote humorous essays on becoming a wheat farmer, janitor or sewer worker—earning me more low grades.

My brother Dave and I got a lot of training as wheat farmers. When we weren't launching rockets on our farm, Dad had us working the fields, year after year. Most of the time it was under a blazing hot sun with temperatures in the upper nineties, breathing in dust laden with traces of radionuclides from the A-bomb tests.

Sometimes we would work all night long when cool breezes would move across the land. The three-bottom plow cut a swath eight inches deep and three feet wide, turning over corks of rich black earth. It took hours to make a single round of the large field, so at first progress seemed hopeless. Three such rounds would reduce the field by only nine feet. But over many days and nights it taught me two important qualities: patience and perseverance. On clear nights I would watch the stars wheeling overhead as they circled Polaris, the pole star.

One night, on our 1935 "Johnny Pop" two-cylinder tractor, I was running low on gas at around 3:00 am, when I noticed a coyote illuminated in the headlight beam. A few moments later, I could make out a half-dozen coyotes circling me and getting ever closer. My heart raced as I reached beneath my seat to grab the two-foot-long wrench for working on the plow. If the tractor died, its light would go out and I would be fending off the wild animals in darkness. But after a while I made out the first slivers of dawn in the east, and to my great relief the coyotes were gone.

The early evening of June 11, 1957 was an exciting time for our science club. We had read in the Wichita Eagle that a large rocket had been launched from Cape Canaveral. The rocket was rumored to be an Atlas intercontinental ballistic missile (ICBM). Such a missile would be able to hurl a nuclear warhead in a ballistic arc thousands of miles over the North Pole to attack enemy targets. Unfortunately, the launch was a failure, with the flight lasting less than a minute. It was assumed by everyone that the Soviet Union was rapidly developing its own ICBMs which could strike the U.S. about twenty minutes after launching. This capability would render delivery by bombers obsolete.

Phil, Steve, and I—calling ourselves the "Mad Three" to separate and distinguish us from the rest of the "proletariat"—attended East High School in the 10th grade, a bus ride from home. There, we were told we needed to stay in place in the school building if an attack was imminent. Nonsense. We planned our escape from each of our classrooms. We would run a quarter mile to the canal drainage ditch where there was a large storm drain that would hopefully keep us from being blown to bits. From my second-floor biology classroom, my escape would be by jumping to a nearby pine tree.

The second launch of the Atlas rocket did not happen until September 25. The rocket exploded after 80 seconds of flight, as reported in the evening TV news. After this, the Atlas rockets began to be covered as live television events which showed large crowds on hand in Cocoa Beach, Florida. Our science club gathered excitedly in my basement on December 17 to watch the third launch of the Atlas. This time, it lifted off and successfully flew 500 miles downrange, to the cheers of everyone. This successful flight was extremely important since the Soviet Union had successfully test flown its first ICBM, the R7, to launch Sputnik into orbit on October 4, portending the Space Age.

We regularly met to watch as many launches as possible, occasionally cutting school. The next Atlas launch on January 10, 1958, was also a success, followed by a row of three disastrous failures. The next success wasn't until June. We followed developments into 1959, when a string of five Atlases exploded. We began to feel deflated, wondering if the U.S. could ever get anything right, after an impressive series of achievements by the Soviets. We began to wonder whether the Soviets would use this opportunity to take a lethal "first strike" at America while we were essentially defenseless. This possibility engendered real fear in all of us.

By late 1961, the government had finally abandoned "duck and cover" as an essentially hopeless idea and began to emphasize fallout protection for citizens far enough from the blast zones to somehow survive. The Department of Defense (DoD) put out a booklet titled, "Fallout Protection, What to Know and Do about Nuclear Attack," undersigned by Secretary of Defense Robert S. McNamara.

The default nuclear yield assumed in the booklet is a 5-megaton burst, in other words, 5 million tons of TNT, more than 300 times higher than the Hiroshima bomb. The booklet explains *"A ground level burst would destroy most buildings two miles from the point of explosion. The destruction five miles away would be less severe, but fires and early fallout could be a significant hazard. At ten miles, sturdy buildings would remain intact. At this distance fires probably would not be started by the fireball, but might be started by the blast wave which could rupture gas lines and short-circuit wires. Flying glass would present a major danger, as would early fallout. At fifty miles, all buildings would remain standing. The fading blast wave would take about five minutes to arrive, but would still shatter many windows."*

Wichita's aircraft industry, especially Boeing and McConnell AFB, was presumed to be a major target. Living less than four miles away from a potential ground zero, I could perhaps survive such an attack by a few minutes, assuming I happened to be in my basement or in a storm drain at the time. Our house was one of the few on our block to have a basement. Mother got a yellow and black nuclear fallout sign from the local civil defense authority and placed it in a bedroom window, signifying that neighbors could seek shelter with us. On the other hand, if I was at our farm, 20 miles southeast as the blast flies, I might possibly have a better chance of survival.

One evening at dinnertime, my father announced, "I'm thinking we should build a fallout shelter." As an architect, he had been attending an evening class on fallout shelter design at Wichita University. "We could build out the basement under the back porch. This would give us about a hundred square feet of shelter space we could close off from the rest of the basement and stock with provisions."

After some thought, I said, "I think that's a good idea." Mother and siblings said nothing.

"Take a look at this," he said, as he handed me a plastic slide rule computer he had received as part of his course. The handy "Nuclear Bomb Effects Computer," was basically a circular slide rule allowing you to input the bomb megatonnage, choice of air or surface burst, and distance from ground zero to calculate the bomb effects.

I spent many hours fooling around with this harbinger of Armageddon. Of particular interest are the blast effects on a person. For example, I could estimate how fast the blast would propel me through the air by rotating the plastic slide rule disks to various settings. For an air burst of a mere 100 kilotons 1.6 miles away a 165-pound man would acquire a speed of 30 feet per second ten seconds after being hit by the blast wave. Double-strength window glass facing the blast would be moving 320 feet per second. I knew death would come in milliseconds.

6 Flying Machines

MY FATHER made countless trips out to western Kansas, supervising the construction of the school buildings he designed. Often, he was gone for several days, so we anticipated his return. We would run out to greet him and check the Buick's grille for birds and other creatures that had met hapless ends. The windshield was usually spattered with grasshoppers and myriad other insects.

Dad was the youngest of seven children; he was a student in the Happy Corner one-room school near Salina, Kansas. He received a set of drawing instruments as a graduation present, inspiring him to earn a degree in architecture from Kansas State Agricultural College (now Kansas State University) after nine years of part time work. He washed dishes, cleaned out butter vats, and sold scavenged potatoes to the poorer people living in the fringes of Manhattan, Kansas. For a while, he had a side business raising skunks and selling their pelts. He was the first in the family to graduate from college. Dad married Virginia Ostlund and was hired by an architectural firm in Hutchinson. That's where I was born. Later, Father had his own successful firm. He was a strong spiritual man, a bit stocky with wavy dark hair and a ready smile; the hardest working person I have ever known.

One time my father returned from one of his many trips out west. I saw that the car, inside and out, and my father as well, was covered by a choking quarter-inch layer of dust from one of the dust storms, perhaps a lingering harbinger from the 1930s.

Looking at me in pained disgust, he told me, "I've had it. I can't do this anymore. I'm getting an airplane so I can fly over this stuff." True to his word, Dad bought a used Piper PA-22 Tri-Pacer and commenced to take flying lessons at Ken-Mar Airport northeast of Wichita. Thus began my further personal evolution in flying machines.

The Tri-Pacer was a four-seater, green and gray affair made of steel tubing covered with doped fabric and powered by a four-cylinder Lycoming engine. It topped out at 122 miles per hour. I was awed and thrilled. Dad's flight instructor was a man named George, a slight and wiry character. Dad took lessons week after week, until he was at last ready to solo. My brother Dave and I were at the airport to witness this event. Mom stayed home.

George took off with Dad for a final ride. After landing, they taxied to the middle of the field, and George got out. Dad taxied back to the start of the grass runway and took off with a roar from the Lycoming and whine from the Sensenich propeller. He made it around the left-hand pattern downwind, base, and approach, but he was too high when he crossed over the power lines, so wisely gunned it for another go around. The next time, he was spot on and made a good landing. Taxiing up to the apron, he and George were all smiles when they entered the building. But the front of my father's shirt was soaked with sweat from his neck down to his waist! On his first attempt at landing, he had failed to account for the lighter weight with George missing from the cockpit.

Dad took many more lessons from George before heading out on his own to western Kansas. Often, no airports were near his jobs, so it was necessary to fly in and out of pastures. This required gaining more experience in crosswind landings, side slipping, and landing with the brakes on. On return flights he would call in with his ETA, and make a few wing-waving passes over our house on his way back to Ken-Mar.

Soon, Dave and I became eager passengers experiencing the magic of flight. Sometimes, Ronnie Gallop would join us as we scrubbed the wings with soap and water to clean off the poop from nesting hangar birds. Dad told us he could get two or three mph better speed with clean wings because of the reduced drag. Our reward was a half hour of flying around with touch-and-go practice.

Flying in the Wichita area airspace required constant vigilance to stay clear of the B-47s and B-52s, bizjets, airliners, and other private aircraft all around. The Ken-Mar runway was almost directly in line with and about seven miles away from the main McConnell Air Force Base runways. Once when we were flying east of Wichita I noticed a small black line smack on the horizon, surrounded by what looked to be an ominous looking little dark cloud. I yelled, "Dad, look out. It's a B-47!" We quickly turned and so did the B-47, but it was a scarily close call.

Dad had joined the informal fraternity of business-men flyers at Ken-Mar. Many flew Tri-Pacers, but Luscombes, Mooneys, and a few twin-engine Piper Apaches populated the small airfield. Once, William Piper himself flew in, and Dad took a ride with him in his Aztec.

Some of the more colorful figures I met were the local TV personality and country music singer Kansas Mack (Sanders) and his corny sidekick Herkimer P. Pushbroom (Joe B. Johnson). They were on TV mostly in mid-day, around the time of the Howdy Doody show. At the airport, I occasionally heard Mack saying no one ever appreciated his music, and how hard it was to please everyone. Tough talking off-screen, he complained that he would likely never achieve success.

One evening, Dad announced we were going to do some night flying. We arrived at Ken-Mar amongst a group of other flyers, including instructor George, Mack and Herkimer. We were planning to do some formation flying under a full moon, in the beautifully clear evening. George organized us into two groups of ten airplanes each. He was heading up the red flight; we were on the blue flight, with my brother Dave and me in the back seat, and none other than Herkimer P. Pushbroom flying with us as co-pilot.

Dad had never flown at night before, nor had he attempted formation flying. George addressed the assembled pilots with the advice, "Pay attention to the plane in front of you and don't worry about who is in back of you."

The red flight took off, one by one, and finally it was our turn. Dad gunned it down the runway, and soon we were aloft. The twenty aircraft were mostly Tri-Pacers. From my position in the left rear seat, the silhouette of a Tri-Pacer appeared hanging magically in the darkness about a hundred feet to the left, and slightly behind us. Another Tri-Pacer was hanging on the right, ahead of us.

I listened to the radio with great excitement as the pilots chatted back and forth with messages like, "Red leader calling red five, over."

"Red five here, over."

"Red five, you are falling behind. Pick up the pace."

Herkimer came on with some of his hilarious TV entertainment personality.

As the mega-formation plowed ahead to the northwest, I was frightfully confronted by the illusion that the airplane behind us had turned and was heading straight for us. I braced for impact and could imagine the propeller cutting into us, both planes going down in a horrific ball of fire. It took more than a few seconds, and some strong willpower, to shake off those feelings of panic.

We were approaching the town of McPherson, about 50 miles from Wichita. George came on to tell us to fly single file over the downtown area, aiming to scare the wits out of the populace. We were successful. At less than a thousand feet above the ground, people ran out of their houses into the darkness in response to our aerial uproar. For good measure, George commanded us to circle around for a second pass, to simulate a forty-plane attack, before setting down at the airport for coffee and snacks.

* * *

Dad said with some urgency, "Look for a place to land! I have to pee real bad." We were halfway to Garden City, several thousand feet above the Kansas flatlands. I was in the co-pilot seat maintaining compass heading, and Dave was sitting in the rear. Dad had taken us out of school for this trip to inspect the construction of a second-floor gymnasium at the Catholic high school. "See if you can find a pasture with no ditches or fences."

Dad took over flying as my vision scanned the checkerboard of square patches below and ahead of us. After a minute or two, I said, "I don't know, how about that one? Just beyond, and to the right of the green one."

Dad banked to the right and cut power, starting his mental landing checklist, calling out, "Carburetor heat on," and so forth.

We made a bumpy landing, rolling up to the barb wire fence. Jumping out, he took care of business with a monumental sigh. Dave and I did the same. Before takeoff, we taxied back along the length of the field checking for obstacles. A hedgerow lurked on the other side of the fence. After a knuckle-clenching takeoff roll, we cleared the trees by a few feet.

Greatly relieved and on our way, in a half hour or so I was able to pick up the Garden City VOR located at the airport. VOR navigation, Very High Frequency Omnidirectional Range system, which we simply called "Omni," had been introduced in the early 1950s. I steered the airplane by lining up a needle on the instrument panel.

Over the city, Dad made a few 500-foot passes over the church where the gymnasium addition he had designed was under construction. Soon, the bishop ran out in his flowing black robes. "OK, he sees us!"

We landed at the airport and were greeted by the bishop who was all smiles and chatter as he drove us back to the church. Dave and I wandered through the nave, while Dad and he met in the office to discuss progress. We had never been inside a church like this with a scary wooden statue of Christ nailed to the cross. The bishop was in charge of things spiritual, but Dad, as the architect, was in command of all things physical.

We were so proud, as the construction crews gathered around to hear him talk about this or that problem. Dad gave Dave and me the run of the construction site, telling us we could ignore the "Keep Out" signs as long as we were careful. We played chess, using steel knockouts from electrical boxes as pieces.

Dad personally inspected construction before any concrete was poured. On this trip, he found all of the steel rebars in the second floor had been installed upside down. Also, the copper pipes running through the floor were of sub-standard thickness. He had specified type "K" thickness, and the plumbers had installed the much cheaper type "M" pipe, cleverly rotating the pipe to hide the type markings. Dad brought a small inspection mirror with him to circumvent this kind of cheating.

He told the men, "Rip it all out! I'll be back to check it again." With this command, weeks of work had to be re-done. The pipes were one thing, but if the concrete had been poured the floor likely would have eventually collapsed, with perhaps hundreds of people inside the gymnasium. Dad told me he sometimes had nightmares that his buildings might collapse and kill people.

* * *

We made many other trips to Garden City. "This is my son, Ralph." Dad introduced me to a young, handsome, slightly stocky young man with dark tousled hair, working on a red and white biplane in the Garden City airport hanger. "This is Mr. Hal Krier." He looked up, pulled a rag out of his back pocket, wiped his hands and extended one to me.

"Nice to meet you."

Awestruck, I managed a, "Hi," as we shook hands.

Dad had known Mr. Krier for some time and had chatted with him on occasion during some of his many trips out west. Krier was a master mechanic and airshow pilot who traveled the country performing aerial acrobatics in a highly modified Piper Cub. Turning toward the biplane, he said, "I'm working on my new airplane. It's a Great Lakes trainer. I call it my Great Lakes Special." As we walked around the airplane, he talked about how he and his brother had modified the airframe to make it faster and stronger for doing loops and other stunts. I was transfixed as I listened to him talk about airflows and wing loading, lift and drag forces.

Then, much to my surprise, Mr. Krier gave us an impromptu flying lesson in the Tri-Pacer. He was co-pilot/instructor in the right seat, and I was right behind him. As we flew over the wheat fields, he and Dad talked a lot about flying as we climbed, turned and banked, and sideslipped back to earth.

I had already acquired an intuitive feel for the four forces governing flight. As a kid, I would hold my arm out Dad's car window to experience the wind pushing my arm backward. That's the force of *drag*. The force of *thrust*, supplied by the car engine, is acting directly opposite to the drag force. In an airplane, it is created by the pull of the propeller or the push from a jet engine. By slightly curving my hand into an airfoil or wing shape, my hand would be forced upward by the force of *lift*. Acting in the downward direction, opposite to lift, is the force of *gravity*. If an airplane is moving through the air with constant speed, these four forces, drag, thrust, lift, and gravity all cancel out so there is zero net force acting on the airplane.

* * *

An excited Ron Gallop was at the front door yelling "Let's go! We can take a ride in a T-34!" I jumped in his car, and we were off to Benton airfield—now known as Lloyd Stearman Field—northeast of Wichita and not far from Ken-Mar. Thirty minutes later, we were at the small airport standing in front of a pair of beautiful silver T-34s. The high-performance propeller-driven airplanes were designed by Beech Aircraft as trainers to prepare pilots for flying jets. Unlike the boxy Tri-Pacer, the T-34 had a narrow fuselage, low wings with retractable gear, and transparent canopy. Ronnie was adept at making connections of all sorts and had somehow conned two airman pilots from McConnell to take us for rides.

Ron's pilot said to my pilot, "Don't fly inverted with negative g's. You'll spill some battery acid."

I climbed into the rear seat of the T-34, strapped in, got a pre-flight briefing, and donned headphones. We taxied to the end of the runway, following Ron's T-34. After a run up and check, we slid the canopy closed and were off with a roar into a wonderfully clear blue sky with fluffy white clouds typical of the Kansas summer.

After climbing together to ten thousand feet, knifing through the edges of clouds, we separated and executed a few exhilarating high g-force stomach-turning aerobatic maneuvers, including loops. It was pretty exciting looking straight at the ground through the top of the canopy!

Now it was my turn to fly the plane. I took the stick and tried a few gentle turns and banks, trying to make coordinated turns by exerting the proper force on the rudder pedals. Soon, I was able to make more aggressive turns with light pressure on the sensitive stick. What an experience! But all too soon it was over.

A week later, my pilot was flying the T-34 and descending slowly from a high altitude, coming in for landing. The automatic low speed "gear up" warning horn was sounding incessantly, so he turned it off. Unfortunately, he forgot about this, and skidded to a landing with the gear up, wrapping the prop around the cowling, and probably totaling the plane. I shivered when I learned about this. It seems a few moments of inattention could easily lead to death.

* * *

Every year there was a big airshow at McConnell. Hopping on our bikes, within thirty minutes we were gliding past long rows of cars waiting to get in. Military jets and small aerobatic planes flew above the long concrete runways. We saw Hal Krier perform his famously crisp eight-point hesitation roll maneuver! I wanted to be like Mr. Krier but knew that would never happen.

The biggest highlight was invariably the Thunderbirds, flying their cool looking North American F-100 Super Sabres. On a hot day with heat waves shimmering off the runway, one of the planes did a 360-degree roll on takeoff and bounced back on the runway with a bang and trail of sparks before becoming airborne again.

"Did you see that?" someone said, as the crowd gasped.

The show was delayed a half hour, but the damaged F-100 landed safely. During lulls in the action, we would sneak our way into some of the hangars, climbing into B-52s to check out the controls, and contemplate the hydrogen bombs that could be held in the massive bomb bays.

One time, Hal Krier's *Blue Skies Airshow* came to a long grass field near Rose Hill, several miles east of McConnell. Cars parked for miles along a north-south hedgerow. Three other biplanes besides the Great Lakes Special were chomping at the bit to get airborne. One was a powerful yellow Stearman. Over eight thousand Boeing-Stearmans were built in Wichita in the 1930s and 1940s.

I said hello to Mr. Krier, my hero, standing next to his plane, but he didn't remember me from our introduction in Garden City. As I turned away, my young ego smarted. But he was distracted, engaged with some other people. I should have invoked my father's name, but my shyness prevented me from thinking of this until it was too late.

The small summer air show was all smoke, noise, and grace. The four planes roared to life and began their takeoff rolls with the Stearman leading the way. Its big radial engine coughed a couple of times, and it was having trouble gaining altitude. Barely making it over the hedgerow, it must have made it to a safe landing somewhere, since it didn't reappear for the rest of the show.

After Krier had gone through his precision flying routine, a thin ribbon was stretched across the grass runway that he knifed through flying on edge. Another ribbon was stretched even lower, and on the next pass he cut it flying inverted 20 feet off the ground! I didn't realize it at the time, but Krier had won the Antique Airplane Association aerobatic championships three years in a row—after which the trophy was retired in his name.

The most thrilling aerial act I saw that day was the wing-suited Batman Red Grant. Roy W. "Red" Grant had been jumping out of airplanes since the late 1940s. Outfitted in a red suit—web-like wings stretched between his arms and legs—his act was billed as death-defying. Apparently, according to the announcer, most men who had tried wing suits had bought the farm after winding up in uncontrollable spins. After much anticipation building up, the batman made his jump, smoke streaming from a canister strapped to his foot. I watched in terror as he spiraled toward the pasture. At the last moment he pulled the ripcord and landed triumphantly before an appreciative crowd.

A few years later, at the Benton airfield where I had flown in the T-34, a Hollywood movie crew filmed *The Gypsy Moths*, which featured winged skydivers, starring Burt Lancaster, Deborah Kerr, and Gene Hackman. My science club friend Steve was an extra in the crowd.

* * *

In the late 1940s, the roar of B-29s coming from Boeing was replaced with the roar of jet engines from new B-47 six-engine swept wing nuclear bombers. We had moved to our new house a few blocks farther away, but the incessant 24-hour din continued. The B-47 was a radical departure from previous bombers in the race for dominance over the Soviet Union, given its high subsonic speed thanks to thin flexible wings. However, the advancements made it tricky to take off, fly, and land. The wings left no room for fuel or landing gears, necessitating an unusual bicycle gear arrangement with a fixed takeoff angle of attack. Over two thousand B-47s were produced, mostly at Boeing Wichita. At adjacent

McConnell Air Force Base, crews were being trained at a furious pace with missions flown day and night.

When I was ten years old, two B-47s collided northeast of Wichita, raining bodies and debris over the farmland; narrowly missing some houses. As a 9th-grader walking to Curtis Intermediate School on a clear March morning, I looked northward to witness a large cloud of black smoke from a B-47 that had exploded shortly after takeoff.

Many other crashes happened in the Wichita area and all over Kansas. Two B-47s collided over Forbes AFB near Topeka. A B-47 from Smokey Hill AFB near Salina blew up in mid flight. Some crashed while attempting to land, some caught fire on the ground, some went into uncontrollable spins, and some had wings tear off while practicing "toss" bombing maneuvers. Many planes from Wichita flew off to crash elsewhere in the country. Nearly all of these crashes were hushed up, and newspapers were warned against covering them. Only decades later—through heavily redacted crash reports—the awful toll became known, racking up 203 crashes resulting in 464 deaths. It was only one of the costs of the Cold War.

I didn't know it then, but years later I would take up my own role in this war.

7 Streaks of Fire

S TUNNED AND DISCOURAGED by the failure of all three of my Vulcan rockets, but thrilled by Sputnik, it was time to regroup. My knowledge of rocketry was sorely lacking. I made bus trips to the library nearly every evening, searching through the stacks for any information I could find about rockets. Alas, there wasn't much—although there were many books about airplanes. One thick book, by von Karman, concerned aerodynamic theory. This seemed to be useful but had many pages of undecipherable mathematics. As a consolation, I checked out and devoured many books written by my literary heroes: Heinlein, Bradbury, and Asimov.

I was beginning to have strange feelings in my body. Walking home from school, I noticed sometimes having tingling sensations in my legs and groin. I knew I was going through puberty but didn't know what that meant. Conversations between male classmates at school were flushed with dirty language. I was not a part of that, and everyone in our science club took an oath to refrain from using this kind of language. We figured there should be no cursing in science.

I had little interest in girls and didn't spend much time worrying about sex. I had somehow assimilated the basics of the sex act and had seen dogs "doing it." The wet dreams were a nuisance. I was careful to wash the sticky substance out of my underpants in the bathroom sink, so Mom wouldn't notice, before putting them in the laundry.

* * *

Scientific American was one magazine I spent hours reading through, learning about many things that interested me greatly, especially astronomy and machinery. One evening at the library, I picked up the June 1957 issue, and thumbed to the Amateur Scientist section in the back. I was stunned to find an article titled "About the Activities and the Trials of Amateur Rocket Experimenters." I could hardly believe my eyes when I read:

"Amateur rocketry has a variety of attractions beyond the spectacular but short-lived flight of the rocket. The design and construction of the missile, its launching apparatus and instrumentation pose many intriguing problems. The rocket must be aerodynamically stable both before and after burnout; this problem of stability can be attacked in many ways and invites endless designs and tests. The design of small motors and special propellants can challenge a sophisticated knowledge of mathematics, physics and chemistry. Because of space limitations the instrumentation of small rockets calls for great ingenuity; it should appeal to radio amateurs, particularly those interested in miniaturization. Similarly, amateur photographers always find a ready-made welcome in rocket societies because pictures are needed of take-offs, flight paths, instrumentation records and the apparatus before and after firing. Being a member of a team that coordinates so many fields of interest in carrying a program through to the climax of firing a rocket with instrumentation gives one a feeling of pride in accomplishment with few parallels."

I probably read the paragraph a dozen times. I realized right then that I wanted to have a feeling of pride for myself. So far, all our group had accomplished was failure, failure, and failure!

Most of the article concerned rockets built by the Reaction Missile Research Society and launched at an undisclosed "rural mountain area." The rockets had combustion chambers made from thin-wall steel tubes with machined steel de Laval nozzles. The propellant was a mixture of zinc dust and sulfur. One of the rockets reached an altitude of 1800 feet!

I began to dream of forming our own rocket group. Our science club would morph into what we would eventually call "Propulsion Research."

* * *

The International Geophysical Year (IGY) began July 1, 1957, and extended to the end of 1958. It signaled a brief pause in the ongoing Cold War. The Navy's Vanguard project was part of the U.S. contribution. Our group of young rocketeers, Propulsion Research, was formed in October 1957, weeks after the launching of Sputnik. The principal members included Steve Hawkins, Phil

Roberts, Gary Peyton, Harold Weibe, Ron Gallop, Jon Freeman, Bob Bullock, Dave Hollis, and me. Others would help out from time to time.

We printed up official Propulsion Research letterhead paper and envelopes. Then, we wrote to the Washington D.C. IGY headquarters, announcing we were in business. Soon thereafter, we began to receive all manner of technical and scientific news documents in the mail, which were discussed with great excitement at our weekly meetings.

Sadly, Bullock had discovered girls and soon dropped out of Propulsion Research. Instead, he focused most of his energy on one girl who aspired to be an opera singer. His younger brother Bruce joined us for some launches.

Shakespeare's famous soliloquy, *To be or not to be, that is the question*, juxtaposes life and death. Should I continue living, trying to defeat the *slings and arrows of outrageous fortune*, or kill myself to enter the sleep of death? I have never entertained the notion of committing suicide. Instead, I interpreted the question differently. *To be*: to strive forward to pursue my passion even in difficulty. *Not to be*: to sit back and enjoy life with a girlfriend or other silly everyday pursuits. Thus, to me, Bullock's defection from Propulsion Research was particularly irksome.

Spurred on by the launch of Sputnik and the discovery of the Scientific American article, we began to reconsider all the problems encountered with the Vulcan rockets. I began to experiment with different fuel mixtures, including zinc dust and sulfur. I added the oxygen-rich compound potassium chlorate to zinc dust, testing its burning qualities. I also tested mixtures of sulfur and potassium chlorate, which burned with a loud roar. I experimented with adding aluminum powder to the mixtures, since I had read in the newspaper that some missiles used powdered aluminum. In the end, I settled on zinc dust mixed with sulfur.

It was clear we needed to take a completely different approach to building and launching our rockets. The rockets described in Scientific American had bodies made of steel tubes, and we hoped that could be the key to solving the problems we had had with the spectacular failures of the Vulcans. At one of our meetings, Phil noted that commonly available CO_2 cartridges were made from steel tubes. The expended cartridges, when filled with a zinc dust and sulfur mixture, might make excellent rocket motors.

I worked furiously during October and November, building a small rocket using a CO_2 cartridge for its combustion chamber. I had previously learned an important principle about rockets: the center of pressure must be located aft of the center of mass. Otherwise, the rocket would go out of control. The heavy steel cartridge at the bottom of the rocket moved the center of mass too far aft. I could solve this problem by adding weight to the nose, but the weight would cut into performance, severely restricting the maximum altitude. Instead, I hit on the idea of using three fins sweeping backward and outward from the bottom of the rocket, hopefully moving the center of pressure aft of the center of mass.

The rocket body itself was a thin aluminum tube into which the CO_2 cartridge, wrapped in a layer of asbestos insulation, was inserted. A six-inch length of piano wire extended from the balsa wood nose cone which, upon its return from the sky, was intended to stick the rocket in the ground. This feature, I reasoned, would prevent the rocket from falling over, which might damage its brightly painted red and white fins. Another improvement was a small aluminum tube glued alongside the body. The rocket would be slid onto a four-foot length of stiff steel wire to give it guidance during takeoff. No more cardboard tubes! I called my new aluminum, steel, asbestos, and balsa wood rocket "Firestreak 1." I figured the Firestreaks would be a series of rockets, so I appended the numeral "1" to the name.

* * *

On November 3, 1957, the world was shocked again by the launch of a second Soviet satellite. Sputnik 2 was a thirteen-foot-high cone-shaped capsule with a base diameter of over six feet weighing around 1100 lbs. It was not designed to separate from the R7's last stage, bringing the total mass in orbit to over seven-and-a-half tons. Moreover, it contained a living animal, the dog Laika, and a rudimentary life support system.

The United States entered a state of raw panic. There were no more arguments about Sputnik 1's weight of 180 lbs. versus Vanguard's proposed satellite weight of 21.5 lbs. It was abundantly clear now that the Soviets had the capability to lob a heavy nuclear warhead right into the heart of America. It was all we talked about at our next meeting of Propulsion Research.

Wichita was a major center of aircraft production, so our fear was renewed that if the Soviets struck first, we were all toast. What we did not realize at the time was that the country would soon embark on a massive effort to catch up on technology of all kinds, and this effort would reach down into the schools and every aspect of civilian life.

* * *

We set up everything at our farm launch site on the chilly morning of November 9, thirty-six days after the failure of Vulcan III and six days after Laika was launched into orbit.

We were reduced to a four-man skeleton crew. Now, some of the members were fed up with spending indeterminate hours in the cold only to see rockets consumed in balls of flame. We set up the main barricade for a photographer and me fifty feet south of the firing area, and a tracking barricade one hundred feet southwest for the other two crew members. A worrisome wind blew from the southwest. Firestreak 1 had been pre-loaded with a mixture of zinc dust and sulfur.

My heart beat faster as I yelled out the countdown so the altitude tracker could hear me. I pushed the launch button. A tremendous whoosh of white flame and smoke exited the rocket as it jumped skyward, ascending to perhaps three feet, pitching upwind in response to a sudden gust and landing fifteen feet away. Finally—a year and eight months after the first Vulcan—a small taste of success! I felt a glow of excitement as I allowed myself a brief bask in this achievement.

The rocket was recovered in pristine condition; the nose spike had done its job! We changed out the combustion chamber and inserted a new pre-loaded CO_2 cartridge. At the end of the day, we had launched Firestreak 1 three more times. One flight reached an estimated altitude of 500 feet, landing 150 feet to the west! Everyone was jubilant as we packed up for the day and headed back to Wichita. We knew we now had the formula for building successful reusable rockets!

A week later, we were at it again with two more launches at the farm. This time we added a remotely operated 8 mm movie camera, moved the cameraman up to a barricade 25 feet away from the launch site, and increased the baseline for the altitude tracker to 200 feet. That day, the best flight reached an altitude of 775 feet, landing 235 feet downwind! This time, the rocket had buried itself in the soft ground, damaging the fins and forcing us to call it a day.

Into December, I continued my propellant and explosive investigations, making fuses out of cotton string soaked in a mixture of potassium chlorate and water. Some samples had added paraffin which, when thoroughly dried and ignited, burned furiously with a violet-colored flame. I also experimented with igniting firecrackers using a hot-wire method and six-volt battery. The explosion of a firecracker might be used to deploy parachutes. All these techniques would be needed for our future rockets.

I worked out a simple algebra formula that could be used to calculate the height a rocket could reach. The formula used the rocket's gross and net weights, the thrust, and the burning time. The latter was short, less than one second for zinc-sulfur, and the thrust was unknown. But by recording the actual height of the rocket's flight with the altitude tracker, it would be possible to work backwards to find the product of thrust and burning time. I thought if we could record the takeoff with Dad's 8 mm movie camera, we might be able to count the frames to find the burning time, and from this determine the thrust. Such was the beginning of our striving to bring a more scientific approach to rocketry.

On December 6, our group met in my basement to watch the launch of Vanguard from Cape Canaveral, carrying a two-pound fourteen-ounce satellite. We gathered in front of the TV set and watched in horror as the beautiful rocket lifted off the pad, fell back, and exploded. The satellite landed nearby and started transmitting signals. For many days afterward, the newspaper and television reports referred to Vanguard as 'Kaputnik.' Experienced as we were with our own failed launches, we all felt terrible on behalf of the scientists.

It was common knowledge that higher altitudes could be reached by dividing a rocket into stages. During Christmas vacation I worked intensely on a 2-stage version of the Firestreak, that I called Firestreak 2. My calculations indicated the first stage would reach a speed of 81 feet per second at burnout at an altitude of 30 feet. But the second stage should reach a speed of 413 feet per second at an altitude of 273 feet, coasting to a height of 2610 feet—almost half a mile!

By early January, I had finished building the rocket. The first stage was a modified version of Firestreak 1 which had a receptacle at the top for the second stage. Fuses for both stages would be ignited together, with a slightly longer fuse for the second stage. Thus, there would be a short delay between burnout of the first stage and ignition of the second stage.

At our next meeting of Propulsion Research, Jon Freeman brought a two-foot length of one and one-eighth inch diameter seamless Chromium-Molybdenum 4130 alloy steel tubing. "My father got this for us. It will make the combustion chamber for a really powerful rocket."

The tube was passed around to the group. I hefted it in my hand. "Much too thick, the wall thickness looks like about an eighth of an inch."

"My dad has a lathe in the basement. I can turn down the wall thickness and make threads at each end."

I was intrigued with the prospect of building a much larger zinc-sulfur rocket. Now we were faced with the task of coming up with a name for this future new class of rockets that would hold about seventeen times more propellant than the Firestreak rockets. I had been thinking of naming new rockets after stars. I knew my way around the constellations and could always recognize the North Star by tracing a line from the pan of the Big Dipper. "Let's call the new rocket Polaris, after the North Star."

This name had met with approval until Hawkins said, "Uh guys, I read the U.S. Navy is building a new type of intermediate range ballistic missile, to be launched from submarines. I think they are calling them 'Polaris missiles,' so we can't use that name."

After more discussion, Freeman suggested the red giant star "Arcturus," which could be found by tracing along the handle of the Big Dipper. It is also the brightest star visible from Kansas. Everyone thought this was a good name. It would take a long and arduous effort to design and build the Arcturus rocket, so for now our focus would remain on the Firestreak 2.

It seemed that people in Wichita had developed a strong interest in rockets. News of our rocket launchings had somehow gotten out to the larger community. On January 15, 1958, a reporter and photographer from the Wichita Beacon showed up at my house. The next day, we were front page news in the evening paper. The headline read "Young Missile Man Ready to Launch 2-Stage Rocket." Over this, in smaller underlined capitals was "NO VANGUARDS ON THIS TEST SITE." The article included a large photo of me in my basement laboratory holding the first and second stages of the Firestreak 2, with Firestreak 1 and the tube of Arcturus standing there on my worktable. The article broke down the team's roles:

"Has Own Missile Team. Each of his friends are assigned a specific task during launching. Stephen Hawkins 'in so many words will push the button and time the rocket flight.' The altitude machine which Ralph built to track the course of the rocket, will be manned by Philip Roberts with Wendell Allen as assistant to the altitude tracker. These three are juniors at Southeast. Gary Peyton, an East High junior, will be the flight timer and observer, and David Hollis, Ralph's brother, will operate the remote movie camera. David is a 9th grader at Curtis Intermediate."

The article concluded by discussing our plans for the Arcturus, which was to be designed for parachute recovery. *"The parachute will be released in the air 'by a small detonating charge, commonly known as a firecracker,' Ralph explained."*

* * *

Two days later, on Saturday morning, January 18, 1958, with news services invited, we set up at the farm to launch both Firestreak 1 and Firestreak 2. This time, we set up at a new location near the farm's south border in a field harrowed and planted in wheat. We had re-designed the launch tower to permit igniting fuses for both stages simultaneously.

My brother Dave, manning an altitude tracker along with Allen, was set up six hundred feet due west of the launch tower near our 1942 Dodge pickup. Roberts, as photographer, was set up forty-five feet southwest, with the main barricade located 30 feet due south, manned by Hawkins and me. A cavalcade of news photographers and TV trucks soon arrived and set up to the south of the tower, with the KTVH-TV truck sixty feet back, and the KAKE-TV truck sixty feet to the southeast. An additional car of random spectators, including my parents for the first time, showed up and parked east of the KAKE-TV truck. The KTVH-TV people strung a cable to a remote microphone near the launch tower. A steady north wind was blowing as preparations were made to launch Firestreak 1.

After Firestreak 1 was slipped onto the launch rail, I called out the countdown. Nothing happened. Repeated attempts over the next hour failed to ignite the fuse. I was sweating in the cold January air as the tension built. Meanwhile, the news people were having a good time walking around and chatting with one another. But they were also worried about meeting their afternoon deadline. Finally, we determined the battery voltage powering the spark coil was too low. Dad, leaping to the rescue, jumped in the truck and headed to the nearby town of Douglass. Forty-five minutes later he returned with a fully charged car battery to save the day.

Eventually, Firestreak 1 leaped from the tower with a loud whoosh and a sweet column of yellow-white flame. The rocket ascended skyward, reaching an altitude of 600 feet, landing 350 feet east of the tower. What a relief! The news people seemed pleased with their photos and video of the takeoff.

It was early afternoon by the time Firestreak 2 with its first and second stages was slipped onto the launch tower. This was the main event, with Firestreak 2's second stage poised to reach a calculated altitude of half a mile! White hot suspense rose as the launch preparations were completed.

I yelled out the countdown. Suddenly, the rocket took off, slower because of the added weight of the second stage. A few milliseconds later, the second stage ignited with a loud roar but instead of going straight up it was cocked at an angle, executed a spiral loop, and crashed into the ground thirty feet northeast of the launch site. The first stage landed forty feet north. That was that. We were done. The news crews packed up and headed back to Wichita.

I suspected the failure had largely been due to the adverse wind conditions and the lack of a guidance rail for the second stage to take off from the first stage.

So much for my equations predicting a height of nearly 2000 feet. Thus continued my education concerning the difference between theory and experiment. Indeed, theories are fragile and delicate things that often contain many simplifications, whereas an experiment, such as actually trying to launch a rocket, is where the "rubber meets the road," as they say. It's the experiment that reveals the truth.

That evening, we were on the KAKE-TV 10:00 news with Greg Gamer, and on KTVH-TV as well. On Sunday, the Wichita Beacon told of the failure with a headline "2-Stage Rocket Fizzles, But Boys Will Try Again." The article had a two-column picture of me in my parka looking dejected while holding the stages of Firestreak 2, one in each hand. The caption read, in part, *"MISSILE MUDDLE—Ralph Hollis, 16, Southeast High missile builder, looks dismayed as he examines the remains of his 2-stage Firestreak rocket."* A nice shot of the Firestreak 1 takeoff was also included. The lead paragraph read, *"Rockets zoomed over 'Cape Douglass' Saturday afternoon, altho (sic) a teenage missile team experienced some of the disappointment of Cape Canaveral."* The article did make note, *"A brisk wind, sometimes sweeping across the flat land at nearly 30 miles an hour, interfered with the ignition of the rockets."*

8 Rocket Science

W E WERE EAGER to move on from our failures. It was time to take a more scientific approach. In late 1958, I had written up some of my equations and Firestreak rocket performance results, failures and successes, and had entered them as a report submitted to the Westinghouse Science Talent Search competition. On January 30, 1959, I received word that I had been selected in the top 10% of the entries; one of four from Kansas. Congratulations came in from Kansas senators Frank Carlson and Andrew Schoeppel. The letter from Senator Schoeppel and his wife said, "We would be most happy to have you drop in at our office. It would be a pleasure to have a visit with you about your scientific studies." That pleased me. A nice article with strange headline "Young Rocket Elected to Honor Group" appeared with my picture in the Eagle. Small articles were also printed in the Beacon and in Stampede, the Southeast High School paper. Only 40 of the best projects would be picked for a free trip to Washington. I was not selected, once again contributing to my feelings of failure.

In the latter part of my high school years, I was consumed by an explosion of interest in many things. I came to love art and poetry, dabbling with abstractions using tempera paint. For a while, I was trying to decide whether to become an artist or a scientist. I started writing short stories, mostly science fiction. I wrote some poetry and met a classmate, Janet Bramel, who also wrote poetry. She had dark hair and wonderfully expressive dark brown eyes. She was a good writer, and I soon realized I was in love with her mind. We became soul mates.

We talked about life, the world, and art. We hung around with Corban Lepell at Wichita University who would later become a well-known artist. Janet and I walked together, sat on park benches, and talked with each other in darkened doorways—whiling away our time in the summer evenings. But mostly we talked about the onrushing future, conversations often tinged with foreboding.

Janet, Steve Hawkins, and I wrote a book of iconoclastic poetry, ran off eighty copies using a spirit duplicator, and sold it in the school hallways. Years later, famous poet Allen Ginsberg, visiting Wichita, learned of our work and was pleased we had the drive and ambition to express our adolescent angst in that manner.

Poetry and art had pulled me one way, but science pulled me back. Inspired by W. Grey Walter's small robots described in 1950 and 1951 issues of Scientific American, Phil Roberts and I built a small robot whose mission was to find its way out of a room by following a wall until reaching a doorway. The robot was self-contained, using small surplus lead-acid batteries for power, and a bank of old telephone relays for logic. Most of the mechanical parts came from Erector sets.

The robot had a single drive motor, but it could turn left or right by jamming a metal stop into either the left or right wheels. Clever differential gearing allowed the robot to pivot around the stopped wheel. Spring-loaded "feelers" on each side of the robot, equipped with small rubber tires, rolled along the wall acting as sensors for the robot's relay brain, completing the cycle of sensing, thinking, and acting. It worked! (Most of the time.)

In the end, science won out over art. But midway through my confused senior year at Southeast High School the seniors were given an "Interest Inventory" test to help determine future career paths. I took the test and learned I was destined to become a playwright.

Representatives from local industry and government agencies were invited to the school, giving talks to the seniors about various careers. We were given a choice of which presentations to attend. I picked one from a Boeing physicist who mentioned little about a career in physics, but instead used his half hour to give us a short lesson in the calculus of integration by parts. I was impressed and chatted with him afterward. He put me in touch with two Boeing engineers, whom I later contacted by phone.

Shortly afterward, I received a small blue booklet in the mail titled, "Pocket Data for Rocket Engines," produced by Bell Aircraft Corporation. I knew, of course, of the famous Bell X-1 rocket plane that Chuck Yeager had flown in 1947, breaking the sound barrier. The booklet was full of equations relating chamber pressure to exhaust velocity, and many other important details concerning rocket flight. This was the information we had been looking for!

* * *

In those days and nights, we were working steadily on completing the Arcturus rocket, and had decided it should have a timing mechanism with a parachute, and a tail cone in lieu of fins to guide it in flight. (The tail cone would circumvent the need to attach fins to the body tube.)

After turning down the diameter of the 4130 steel tube, Jon Freeman struggled on his basement lathe to machine threads on its ends to form the combustion chamber. He machined a threaded de Laval nozzle out of steel to screw into the body, with throat dimensions calculated from a formula given in the booklet. The formula for the throat dimension—the opening's diameter—involved a square root, which Freeman calculated on his slide rule. When I saw the finished nozzle, I thought the opening was much too small which would likely lead to a catastrophic explosion. Square roots have so-called "real" and "imaginary" parts, and he had used the wrong root! A few hours later the throat was bored out, correcting this error, and the transitions between the entry and exit sections were smoothed down.

On advice from his father, Freeman decided to "spin form" the tail cone from a thin disk of aluminum. First, a steel cone with a flat on one end was machined on the lathe to act as a form. Then the disk with a center hole was attached to the flat using a screw. With the lathe spindle spinning rapidly, Freeman forced the end of a broomstick against the spinning disk as I watched it magically begin to flow around the cone, heated by the friction. Suddenly, the flowing stopped, and the aluminum cracked with a bang. We tried this many times with different disks and varied broomstick force, but each time the aluminum cracked. Once again, reality had intervened; something was amiss with our theory.

The next day, Freeman's dad looked at our pitiful results and told us we were using the wrong kind of aluminum; we should switch to a softer alloy. We were able to get the softer kind of aluminum at a store called The Yard. This worked great on the first try. It was so much fun we turned out a half dozen more cones for the heck of it.

51

The sprawling Yard store on Central Avenue was the go-to place for all sorts of aviation surplus items. Stacks of crates containing twelve-cylinder aircraft engines laid out in the open, and a half-dozen Quonset huts chock full of small electric motors, gears, and millions of fasteners offered a smorgasbord of goodies for experimenters. Raw aircraft aluminum tubes, sheets, and plates were stacked up all around—heaven for us rocket builders.

* * *

Eventually, Janet drifted away from me. Jean Taliaferro was another girl in my class. We sometimes walked home together to the corner of Oliver and Morris where we parted, and I would walk the remaining half-mile to my house. I liked Jean. We mostly talked about class assignments. I told her about my friends Phil and Steve, and about our rocket building. I met her family living four houses down from our corner. Her mother was a band teacher, and her father worked at Cessna—bucking rivets on their new light planes. Later, her father would take night classes at Wichita University, studying to become a teacher.

Jean's sister Audine was two years younger, Marjorie was next in line, then Carol, and lastly Howard, a kindergartner. Before long, Audine began to join us on our walks home from school. I liked Audine, and it wasn't long before we started hanging out together. Pretty, with light brown hair and hazel eyes, one of her upper teeth was recessed instead of overlapping her lower teeth, a malocclusion, like my own. We liked drive-in movies and fast-food places. She was interested in my rocket activities and soon became an honorary member of Propulsion Research.

* * *

By this time, Father had purchased an additional tract of land nine miles east of the Hollis farm. It was mostly grassland, so we called it the "Ranch." As our rockets got bigger, we began to think of moving our launches out there, farther from Douglass. We made a nicely painted white wooden sign declaring "Propulsion Research Test Site," in black lettering, erected on a post near the gravel road at the corner of the property. A few weeks later it had been shot full of holes, presumably by the gun-toting locals who regularly shot up stop signs in the area.

We expected our Arcturus rocket to reach at least a thousand feet. So, it was clear that we needed a parachute to lower it gracefully to the ground. Ideally, we wanted the parachute to pop open at the apogee of flight. To time the parachute release, I remembered a Japanese device Bob Bullock's father had brought back after the war—some sort of inactivated grenade. Bob and I had played around with it when we were in grade school. Three disks were stacked one upon another and fastened together with a screw through their centers. Each disk had a circular groove with a small hole in its lower surface.

Bullock's father had explained to us how when the grooves and holes were packed with burnable powder, flame could propagate upward from the bottom-most hole into and around the groove of the bottom disk, and thence to the second disk through its hole and on to the third disk. The bomb would go off when the flame exited the hole in the third, uppermost, disk. By rotating the disks with respect to each other, one could set the time between initial ignition and final explosion over a large range—sort of like an adjustable length fuse. This seemed like a good idea for Arcturus, allowing us to time the parachute release.

Timing the release at the apogee would depend on thrust and burning time. The problem of calculating the rocket thrust gnawed at us. We knew the "big boys" at Cape Canaveral had equipment to measure such things, so in early January 1959, we decided to build a so-called "thrust jack" to measure thrust. The rocket motor, held down, would push against a piston tightly fitted into an oil-filled cylinder. The oil pressure, displayed on a gauge, would be recorded by our 8 mm movie camera, and the motor's thrust would be the pressure times the piston's area. The whole apparatus would mount horizontally on a steel plate welded to a large steel pipe set into the ground with concrete. Easier said than done, but no task was too daunting to stand in our way of taking a more scientific approach to rocketry.

At the Ranch, with most of the group participating, and with help from Dad, we set our welded test stand into a four-foot-deep hole in the middle of the field, leveled it up and secured it with a ton of poured concrete, mixed by hand. We figured it could support horizontal thrust loads of up to at least 10,000 lbs. To test the thrust jack, Audine held the cylinder while I poured in a quart of SAE 50 weight oil. After closing the fill port, I pushed hard on the piston, simulating what the motor would do upon ignition. A thin stream of oil squirted out of the cylinder hitting Audine in the face at her hairline. She jumped back, cursing and soaked with sticky oil. This did not portend well for our future relationship.

Several weeks later, after the concrete had cured, we loaded the Arcturus motor with zinc-sulfur and attached it to the test stand. Unable to seal the thrust jack leak we pressed ahead, sans thrust jack, for the motor's initial test. Freeman, Gallop, Hawkins, and I crouched behind a newly made barrier constructed of old two-by-sixes we scavenged from the remains of a small barn. The movie camera was set up to peer over the top of the barricade, and a large wooden-framed bathroom mirror was set up on the ground at an angle. We could see the firing in the mirror while safely remaining behind the barricade. If the motor blew up, it might take out the mirror, but that would be the cost of research.

As a feeling of dread settled over our small group, Freeman started the movie camera, I yelled out the countdown, and Gallop pushed the button. The engine ignited with a tremendous roar and eight-foot yellow-white exhaust flame. The test was a success! We counted the movie frames to find the burning time, but still hadn't found a way to measure the thrust.

We finished the Arcturus rocket in early summer, complete with parachute release mechanism and tail cone. Unfortunately, the launch was a bitter disappointment. The rocket launched straight up, climbed to an altitude of around seventy-five feet and fell back to the ground. Upon recovery, the tail cone was bent, and broken free from the exhaust nozzle. The thin aluminum parachute container was also broken off at the top of the motor, never having had a chance to deploy the parachute. Once again, it was time to go back to the drawing board.

* * *

We needed a way to calculate the performance of our rockets. Slide rules had helped to determine the burnout velocity and maximum height they would achieve, but we needed a better approach. We had heard of many other amateur groups and individuals who were building rockets. Some were experimenting with exotic propellants, and some of the rockets were exploding with disastrous results. Unbeknownst to us, nascent model rocket companies were forming, most notably Estes Industries near Denver, Colorado. Propulsion Research had become but one group of amateur rocket experimenters.

We needed to distinguish ourselves by building bigger, more advanced, higher performing rockets. We should strive to be the first to achieve flight into the stratosphere. That would put Propulsion Research on the map. To accomplish this ambitious goal, we needed to mathematically analyze our designs. This would make our approach different from all the others.

During my senior year at Southeast High, I had made a few excursions to Wichita University (now Wichita State University). I wandered through the engineering building and around the Walter Beech wind tunnel—cautiously poking around some labs to find out what was going on. Once, I saw an exciting demonstration of a model pulsejet engine that traveled with a roar along a taut wire stretched between buildings. This type of engine was used on the German V1 flying bomb which had caused so much destruction on London.

I found a room designated as the "Computer Lab," according to a sign next to the door. Students were doing their homework using Friden electromechanical calculators, arranged in a row on two long tables. Each person was entering numbers on a sloping panel with ten columns of pushbutton keys, each embossed with digits from one to nine. Pressing these keys would cause the numbers to appear as a row of numbers on a moving carriage—a little like the carriage on a typewriter—above the columns. After a number was entered in this way, a second number could be entered, causing a second row of numbers to appear on the carriage. Operations like multiplication and addition were performed by pressing corresponding keys, after which the carriage would move along with a wonderful clattering and crunching sound until the result, up to twenty digits, would appear on the carriage in a matter of seconds. As I watched in amazement, I realized these calculators were the breakthrough we needed.

We wanted to calculate the flight of a rocket, second by second, from takeoff until the maximum altitude was reached. First, some basic assumptions were considered. The combustion chamber pressure should be as high as possible, subject to its bursting strength. This, in turn, depended on the material—cardboard, aluminum, or steel—as well as its thickness. We knew the rocket's thrust depended on the speed of its exhaust gases which, in turn, depended on the square root of the combustion chamber pressure. The downward momentum of the gases escaping in the exhaust would impart an equal and opposite momentum to the rocket, according to Newton's third law: for every action, there is an equal and opposite reaction. Momentum is the product of the exhaust speed and the mass of the exhaust gases.

We had assumed that the rocket's propellant started burning at the bottom of the combustion chamber, proceeding upward until burnout, like a burning cigarette. Our experiences with the Firestreak and Arcturus rockets had indicated, however, that the burning rate—which, according to the Scientific American article, could be up to ninety inches per second—and hence the thrust, was quite variable. This concerned me as we aspired to build our much bigger Hyper-Force and two-stage Cryolite rockets.

* * *

Six of us snuck into the University's unoccupied computer lab the following weekend. The six Friden model SRW machines were a godsend to us. They could perform multiplication, division, and square root operations. Amazingly, an SRW could extract the square root of a ten-digit number in nine seconds! We needed square roots to calculate nozzle throat areas from combustion chamber pressures. Most importantly, the calculator could perform addition and subtraction, operations we could not do on our slide rules.

Hawkins, Roberts, Gallop, Freeman, Dave, and I were lined up in order, left to right, each in front of his own Friden calculating machine. The immense power of this assemblage was not lost on us. We knew we had to apply the integral calculus to solve our problem, but up to this time, none of us had studied more than basic analytic geometry. I had, in fact, earned no higher than a D in trigonometry.

"Okay, Hawkins, the time is zero. Enter a thrust value and calculate the initial acceleration," I said.

Crunch-a-crunch, "The acceleration is *such and such*," Hawkins said, knowing the initial weight of the rocket and its propellant.

The next man, Roberts, would multiply this result by the elapsed time of one tenth of a second, and add it to the previous speed, (zero at this point), *crunch-a-crunch*, before calling out, "speed is *so and so*."

Then, Gallop would use Roberts' speed value to perform another multiplication and addition to find the altitude reached in the first tenth of a second.

Since the imaginary rocket was now traveling upward with some speed, it was encountering air resistance, or drag. The drag depended on the square of the speed and the shape of the rocket; the latter was estimated from formulas I had found in von Karman's book.

So, Freeman next calculated how much the drag was, *crunch-a-crunch*, and called out "drag force is *such and such*."

Then, Dave would use Freeman's drag number to decrease the effective thrust needed for the next tenth of a second, since the drag acted directly opposite the thrust force. However, once that tenth of a second had passed, I had to calculate the new, lighter, weight of the rocket, since some of the fuel was now expended. Now the ball was back in Hawkins' court. I would say, "Hawkins, the weight is now *such and such*, calculate the acceleration for the next tenth of a second."

At each stage of the calculation, when each man called out his number, I would write it in my notebook, gradually building up a profile of the imaginary rocket's vertical flight. Each tenth of a second took only a few minutes to compute using the powerful machines. After all the imaginary propellant was expended, we switched to one second time intervals to complete the remainder of the flight. Later, I would plot the results on graph paper. This iterative human-calculator mega-machine would operate for many hours until we completed all the calculations after which, exhausted, we all went home.

* * *

One day in late April 1959, I wandered into the engineering building at the University to find an amazing sight. In a small glassed-in room built over a stairwell, I saw a big gray machine. It was six feet wide, four feet tall, and two-and-a-half feet deep. The top had a hinged lid that could be opened to reveal some unfamiliar mechanisms holding reels of paper tape. On the right, a small table held a gray typewriter, and some sort of console with switches and buttons. Two men were in the room talking about the machine. Curious, I entered the room and asked, "What is it?"

One of the men said, "It's a computer. An IBM 610."

Later in the day, I came back to investigate. A man, maybe a professor, was typing something on the console keyboard causing a mechanical sound coming from the top of the machine. I asked him what he was doing.

"I'm punching a program on the tape," he said.

"How does the computer work?"

"The manual is over there," he said. The 65-page Operator's Manual was attached to the computer by a cord tied around a hole punched in its corner. I sat there reading the manual for many hours trying to understand how to program the machine. I learned the computer had over 2,000 digits of memory, stored on a rotating magnetic drum. It had 84 special memory locations called "registers," of 31 digits, each of which could be used as a so-called "accumulator," where intermediate arithmetic and logic results could be stored. In addition, 6,000 digits of intermediate, automatically re-circulating memory was available through the punched paper tape unit.

Several different operations could be performed on the 610, including addition, subtraction, multiplication, and division, as well as square roots, pretty much like the Friden SRW calculators. Also, left and right shifting, and other operations for reading and writing data were provided. But, unlike those operations on the electromechanical calculators, there was no need to yell results back and forth between multiple human operators. With the 610, everything could be done within a single machine, controlled automatically by a program. The implications were thrilling.

For days afterward, I returned to the 610 and talked with some people who were constructing programs to run on it. At last, I asked the person in charge if I could use the 610 to make rocket calculations. Permission was granted, even though I had not yet graduated from high school. "Just reserve time in the logbook and sign in with your start and ending times."

I learned that the machine could be operated manually from the keyboard, a bit like the Friden. But it could also be operated by a program punched into paper tape. The pale-yellow tape, about an inch wide, had eight rows of punched holes, plus a row of small holes that engaged a sprocket to move the tape along. The machine also had a removable board, or patch panel, covered with small holes or sockets, where short cables could be plugged in, connecting pairs of sockets. Evidently, the computer could also be operated using the patch panel. Results from any calculations carried out could be printed on the automatic typewriter or punched onto tape. I was filled with mystery and excitement as I began to contemplate how I might be able to use the 610 for my rocket calculations. How in the world could I figure out how to make a program that would actually cause the machine to calculate something I needed?

Gradually, I began to realize I should analyze the problem to be solved, breaking it into a sequence of basic operations expressed in the 610 language (a cryptic bunch of numbers), and punched onto the program tape, or else wired on the patch panel.

I had to decide what input data should be entered onto the data tape, and what output results were to be typed on the typewriter, and in what format. I also had to decide what storage locations should be used for the various constants and variables needed.

Finally, I would need to write out, line by line, the precise instructions that would input data into the 610's memory. I was intrigued by the patch panel, which could be removed from the computer and wired up. Several of these panels were available and could be reserved. If I needed to use the panel, I would need to concoct a diagram for wiring it. A convenient stack of printed forms helped to plan the wiring by using a pencil to draw all the connections.

Soon, I began to try many different calculations. Not only could I calculate rocket trajectories, but I also began to calculate gas flow effects inside my rocket motors, so-called "internal ballistics." For example, I could calculate the theoretical velocity of my rocket exhaust as a function of the gas pressure and temperature inside the combustion chamber, given the specific heat ratio of the propellants. The specific heat is the amount of heat required to raise one lb. of the propellant by one degree Fahrenheit.

In a rocket motor with a de Laval nozzle, the gas is subject to a range of conditions as it expands and exits the nozzle. Knowing the rocket exhaust velocity is important, since the thrust force is directly proportional to the exhaust velocity. The calculation involved a number of ratios involving the so-called universal gas constant and combustion temperature. After calculating several terms and an exponential as a sub-program on the patch panel, a final square root operation yielded the results, printed out on the typewriter for later graphing. Having the 610 at my fingertips was a godsend.

* * *

It was evening. Audine and I had walked together back from my parents' house to hers. We paused for a few minutes under the streetlight at the corner of East Morris and Oliver. I drew her in close to me; our lips coming together slowly, hearts pounding, we shared a soft, wondrous kiss—the first and most memorable kiss in my life. I thought we must be in love.

9 Hyper-Force

T HE DEADLINE for the Wichita High School Science Fair was rapidly approaching. It was my last chance to present my rocket work to the general public before graduation. At Propulsion Research we had been thinking of designing a much larger rocket we called "Hyper-Force." The combustion chamber would be six inches in diameter and nearly forty-eight inches long. It was based on a steel oxygen tank purchased from The Yard which, we imagined might have been salvaged from a crashed B-47 or B-52.

The Hyper-Force design incorporated a four-inch diameter upper section made from an aluminum tube, containing the Parachute Timing Release Computer (PTRC), designed by Harold Wiebe a year before. Harold had been working on the PTRC as an electronic version of our earlier electromechanical design. A large parachute, to be ejected by a firecracker explosion from the side of the upper section, would be used to recover the vehicle, a feat that had so far eluded us.

Calculations, computed with the help of the 610, indicated Hyper-Force would have a ten-to-fifteen-mile altitude capability. We were using design formulas given to us by two engineers from Boeing-Wichita, as well as formulas from the small Bell Aircraft booklet.

The weekly meeting of Propulsion Research in my parents' basement was called to order by Steve Hawkins, secretary. After the previous meeting's minutes were read and approved, attention turned to Harold Wiebe. "Guys, here's my idea for an *integrating accelerometer* that will measure the rocket's acceleration and calculate its velocity using an analog electric circuit," said

Harold, illustrating with his arms. "After launch, when the velocity reaches zero at the apogee, it will cause a signal to ignite the firecracker." He laid out several sketches in front of us. "Instead of using vacuum tubes, I am designing it using transistors." We had read about transistors in Popular Mechanics magazine and were eager to hear more.

With a flourish, Harold pulled a small board out of a box he brought to the meeting. "My test circuit—this is a CK-722, made by Raytheon." He was pointing to a small blue thing the size of my pinkie nail. "At several dollars apiece, it's kind of expensive. But I'm thinking it may be the future." With this prognostication, he connected a wire to the battery. A small puff of blue smoke emanated from his CK-722, to everyone's surprise and laughter. That is, everyone but Harold.

"Hey, Ralph," announced Phil. "They're testing a big solid propellant rocket at Edwards Air Force Base in California. It's called 'Minuteman.' They're launching it from an underground tube with a string attached to make it fly in an arc and crash into the ground. I guess they don't want it to fly very far." He chuckled about what a clever idea that was. "I read about it in my dad's *Aviation Week* magazine." Phil's dad was an engineer at Boeing. Years later, I would need to become intimately familiar with the third generation of this missile.

We heard information was available from the U. S. Army Artillery and Missile School at Fort Sill, Oklahoma, 250 miles southwest of Wichita. Writing to Fort Sill on the Propulsion Research letterhead, we soon received "A Guide to Amateur Rocketry," compiled in early 1958. The booklet was amazing. The dark yellow cover showed a drawing of a rocket taking off and climbing into the sky. The back cover showed some rockets ready to launch, with the captions "Preparing to Fire," and "The Army Assists."

As I learned many years later, the booklet was an aftereffect of the National Defense Education Act (NDEA) of 1958, a stunned reaction to the Soviet Union's surprise launch of Sputnik. Incredible as it may seem today, this guide to amateur rocketry was issued with the blessing of Commanding Major General T. E. de Shazo.

The NDEA was signed into law on September 2, 1958, providing funding to education institutions at all levels. It followed a growing national sense that American scientists were falling behind those in the Soviet Union. Title V of the law included provisions for the implementation of testing programs to identify gifted and talented students. Toward the end of my senior year, I was one of many taking national tests in chemistry and physics.

* * *

It was the day of the High School Science Fair, held at Wichita University. As a senior, it was my last chance to participate in a science fair, but I had run out of time to build and launch the Hyper-Force rocket. Instead, I used the surplus oxygen tank and other parts to build what was essentially a full-scale non-operational model. My brother Dave helped with putting things together at the last moment, using epoxy and Bondo auto body compound. Later on, I could re-use some of the model parts to build a real Hyper-Force.

My poster illustrated many different performance calculations for the Hyper-Force made using the IBM 610. These included the effect of propellant burning rate on burnout velocity, combustion chamber diameter effect on burnout velocity with various payload weights, and so-called thrust coefficient as a function of chamber pressure, among others. I had carefully used the 610's typewriter outputs to make graphs using pen and ink. All of these graphs and photographs of our launch site and my crew members were glued to a large cardboard poster. The Hyper-Force model stood next to the poster. I had also written a brief research paper describing my efforts.

When the panel of judges eventually reached me, I suddenly realized I was supposed to make a short speech describing the work. Standing there next to the Hyper-Force model, I was gripped in a wave of panic as the judges took a minute or two to scan the various graphs, then looked at me, expecting me to say something. Terrified, my throat choked up as I was about to speak when the judges turned and walked off.

A few moments later, a newspaper reporter asked if he could take a picture of me next to the rocket. I was beginning to tire of all the press attention, so I told him I wasn't interested. The following day, a short article appeared in the Beacon:

"Ralph Hollis, 17, who is a senior at Wichita Southeast High School, exhibited a model rocket at the Science Fair at the University of Wichita.

The rocket was about twice his size, and when photographers wanted him to pose with his model, he refused. Hollis explained that he wasn't in this business of rocketry and science just for fun or publicity.

`I'm a scientist,' he explained. And he didn't have his photo taken either." When it was all over at the end of the day, my project was awarded third place out of perhaps a hundred entries. This did not make me especially happy.

* * *

America's first space satellite, Explorer I, was launched January 31, 1958. At last, a version of Singer's MOUSE that inspired my science project in 1955 had actually come true! Three months after the historic launch, Dr. Wernher von Braun arrived in Wichita as part of his nationwide victory tour. His rocket had launched the satellite. Hawkins, Roberts, and I arrived early to grab seats on the first row in the Wichita University fieldhouse, only fifteen feet from the lectern.

The great man walked in behind an ROTC honor guard, dressed in a crisp fitted suit and tie, looking the essence of authority. After a brief introduction, he began to speak in careful, exacting tones, as we sat there entranced by every word. Here was the man himself, in the flesh. Here was the man behind the Collier's magazine visions of interplanetary space travel. He spoke for only half an hour, after which he was spirited away with no questions permitted. It was simply stunning! Hugely inspiring and something I will never forget.

* * *

Graduating from Southeast High School in 1959, my grade point average was 2.95, better than a C, but not quite a B. That summer, I worked again at Reiss and Goodness Engineers, as I had the previous two summers designing sewer systems for small towns, mostly in western Kansas. The work was five days a week plus Saturday morning, and I was expected to sweep out the office every morning before 8:00 am.

Jon Freeman had joined me there. Phil Roberts had a job at a company testing the strength of concrete, and Steve Hawkins had another job at a company working as a drill press operator.

Sometimes, I would go on trips to western Kansas with the aggressive foreman, Mr. Jollie. He would stand in the shade of a tree using a theodolite instrument, while I would walk all around in the oppressive heat holding a tall stick marked off in feet and inches. He would wave me this way and that, yelling at me to wade waist deep into murky, leech-filled swamps. These lower elevations marked the location of future sewage treatment plants. At each location of the stick, he would write down the sighted elevation in a little book.

On occasion, we shared a room for the night in some dusty hotel, where he would regale me with complaints about the company's management and life in general. On one memorable trip, I had to measure the basement depths of all the houses in a small town. In one of the basements, colorfully dressed ladies fussed over me while I was taking readings. Much later, I realized I had happened into a "house of ill repute."

Back at the office, I would use Jollie's notes to draw contour maps by hand to visualize the lay of the land. I would plot the readings at all the locations on the paper, then draw in the contour lines, performing the interpolations by sight. Using the basic principle of "shit flows downhill," I would then lay out the paths of sewer lines branching from the lowest elevation to the various streets in the town.

I began to wonder if I might be able to calculate the contour lines more accurately by using the 610. I successfully programed the necessary two-dimensional interpolations while working at night. I mentioned this to my boss, Mr. Goodness, who was unimpressed. Nevertheless, I began to realize computers might be able to do routine calculations like laying out sewer lines. I mentioned this notion to my father, suggesting he invest $500 in IBM. Alas, he never did. In those days, throwing $500 in the stock market was not something one did lightly.

Once, I was told to do some surveying with the hot-headed Mr. Reiss, company president. He was running the theodolite and I was out a few hundred yards away along a roadway holding the stick. I was probably daydreaming about rockets when I looked up and he was waving and yelling at me. Each time I moved the stick as I guessed what he was trying to tell me, he grew more animated. Finally, he threw up his hands, jumped in his car, and accelerated directly toward me reaching an impressive speed. *He's going to kill me!* At the last moment, I jumped out of the way as he zoomed past me and skidded to a halt in a ditch. After he cursed me out, I swore I was quitting. But I swallowed my pride and stayed on because I needed the $300 summer's pay to finance rocket building.

Leaning over my drafting table at Reiss and Goodness, I daydreamed a lot about the relationship between science and math. It seemed to me that mathematics represents the pinnacle of knowledge; physics was next; then engineering with its practical concerns. I decided to start by studying mathematics. If that didn't work out, I felt I could drop down to studying physics. If I couldn't hack physics, there was always engineering. That was my plan.

10 Cryolite

A UDINE AND I WERE STUDYING in her parents' living room. "Well, good luck, honey," said Audine, looking up from the dozens of flash cards she had made for quizzing me on the trig identity formulas. I had studied the book all summer and worked all the problems, cover to cover, in hopes of quizzing out of the University of Wichita algebra and trigonometry requirements. The fact that I had received a mere D in high school trig was a big problem. The next day I would start my major in mathematics.

"You're terrific. I'll do my best." Rather than move to a campus dorm, I would stay at home with my parents and siblings; a convenient and cost-saving arrangement. Indeed, many students attending Wichita University lived with their parents, considering it—as I did—only an extension of high school.

Alas, although I remained in Wichita, Propulsion Research was beginning to disintegrate. Harold Wiebe had moved to Cincinnati before his senior year at Southeast. Phil Roberts had started his freshman year at the University of Kansas in Lawrence, and Jon Freemen had started his freshman year at the University of Colorado in Boulder. These friends were heading off in all directions, but I still had Gary Peyton and Steve Hawkins with me at the University of Wichita as well as Ron Gallop, my brother Dave, and Audine, who were all high school juniors.

It was late afternoon when I handed in my tests, waiting expectantly in the empty classroom for the verdict. Thirty minutes later, the math instructor emerged from his office. "You missed a few, but I'll mark you an A in algebra. You had a perfect score in trigonometry."

"Okay, thanks," I said, smiling meekly and feeling greatly relieved.
That evening, I took Audine to the movies. "Well?" she asked.

"I'll tell you after the movie." The movie was pretty boring. As we walked out of the theater, I told her my results. She jumped up and down, squealing in delight.

* * *

I was purposely exposing myself (or subjecting myself?) to the mysteries of advanced mathematics, but I could not take my mind off rockets. Cryolite is an ore used in the production of aluminum. It would be a good name for the giant two-stage rocket we were envisioning. Despite the public failure of our two-stage Firestreak, we knew multiple stage rockets were the key to reaching extreme altitudes. Rather than rebuilding the Hyper-Force model into a working rocket, we headed in a new direction. It was clear that if we could make our rockets lighter, they would fly higher. Thus, we figured they should be made from lightweight aluminum, which we could get from The Yard.

As a test, we started with a small rocket whose combustion chamber would be machined from a solid cylinder of aircraft grade aluminum. Ron Gallop would do the machine work. The rocket would be 100% aluminum with a welded de Laval nozzle and three double delta fins welded onto the body. I had used the IBM 610 to calculate there would be 87 lbs. of thrust at a chamber pressure of 1000 lbs. per square inch.

On April 10, 1960, we set up for the launch at the farm. This time, we did not invite the press. We were met with a clear blue sky and warmer temperatures as preparations were ready at 3:00 in the afternoon. We tried a resistance-type igniter for the first time, which had a small coil of nichrome wire anchored with a rubber stopper wedged in the rocket's nozzle. Gallop observed from 400 feet south of the launch site and I observed from 350 feet southwest. Dave pushed the button at a point some thirty feet from the rocket, protected by my car.

After a two-second delay, the rocket screamed into the air on its maiden flight, accelerating at an almost unbelievable rate. I saw the rocket at launch, then roughly 500 feet in the air, then several glitters in the sun at the thousand-foot level. Several seconds later, it plowed in, hundreds of feet north of the launch site and was recovered completely undamaged. The flight was in all respects perfect and completely stable. "Very promising!" I said to Ron.

"It has four times the propellant compared to the Firestreaks," he said, with a big grin. We reloaded the combustion chamber with the zinc-sulfur propellant; after a four-second delay the rocket again leapt into the sky, perfectly straight and true. I observed the rocket as it rapidly faded into a tiny dot and vanished.

I watched for it to come down, but I couldn't see it anywhere. The rocket must have climbed to 1300-1600 feet! The three of us were thrilled to our bones. After searching for many hours, we gave up trying to recover the rocket, figuring it probably dug its way into the ground. Plowing and harrowing the field for years afterward failed to uncover it.

* * *

With the success of the nameless small all-aluminum rocket, it was time to work hard on our epic two-stage Cryolite rocket, whose diameter was nearly 3 inches, and assembled length was over 12 feet. The entire rocket was made of lightweight surplus 2024 aluminum alloy we got from The Yard. Cryolite's first and second stage combustion chambers had wall thicknesses of one sixteenth of an inch, which would support combustion pressures of 1000 lbs. per square inch—at least at room temperature. We knew the aluminum would weaken as the temperature increased, but with such a short burning time, we hoped it wouldn't be a problem.

The first and second stages would have five and six-foot diameter parachutes, respectively. The parachutes were to be deployed by explosive charges using timing devices following the design of Harold Wiebe's Parachute Timing Release Computer (PTRC).

Ron Gallop was busy designing and building a radio transmitter out of miniature vacuum tubes, to be housed within the carved balsa wood nose cone. The transmitter would aid in finding the second stage if it landed miles away. In the future, it could be used to transmit data from onboard sensors.

Both stages would have a triad of triangular stabilizing fins at their aft ends, as well as conical jackets surrounding the de Laval nozzles. The innovative jackets were to be loaded with titanium tetrachloride liquid for cooling the nozzle throats, which we hoped would reduce erosion. The jackets were to be sealed with wax that would melt from the heat after launch, releasing the titanium tetrachloride liquid which would react with air to produce billowing white clouds of titanium dioxide and hydrogen chloride. This would enable us to track the flight to high altitudes long after burnout.

I worried about the loosely packed zinc-sulfur propellant which would burn in cigarette fashion with a speed of up to ninety inches per second. But what if the takeoff acceleration was so high that the mass of propellant would slide down into the throat, jamming it up and causing a spectacular explosion?

We began to rethink the propellant. What if the sulfur could be melted and the right amount of zinc dust powder could be stirred into the mixture, and then cast into a solid, slower-burning propellant? I made a few tests, adding two parts of zinc dust to one part of molten sulfur, by weight. I cautiously heated the mixture to its melting point of 112 degrees C using a propane torch, mindful it could explode at any instant. After stirring, I poured the molten mixture into a small juice can and set it aside to cool.

The next day, I cut off the metal can surrounding the cast propellant. Out on the back patio, I focused the flame from the torch on the propellant for at least a minute, but it refused to ignite. By putting a small mound of loose zinc-sulfur propellant on top of the cast slug, I was able to achieve ignition with the torch which caused the cast propellant to burn slowly and steadily.

After that experience, I hit upon the idea of casting the propellant inside the combustion chamber with a central hole running down the center. This would be similar to the first Vulcan rocket, where we packed black powder around a plaster cone. Further, I planned to fill the hole with loose zinc-sulfur powder. The powder would burn quickly, providing high takeoff thrust, while also igniting the slower-burning cast propellant which would continue a sustaining thrust until burnout. The cast propellant would burn from the inside out, preventing the flames from reaching the combustion chamber's thin wall until burnout is reached.

A big problem was finding a way to machine all the dozens of parts for Cryolite. We definitely could not afford to hire a machine shop to make these for us. Freeman was gone, so his basement lathe was not an option. It was agonizing. We were stuck in a cycle of despair.

11 Production Machine Company

T HE WINTER OF 1961 was brutal. Western Kansas endured an epic blizzard— the kind where you need to follow a rope tied between the house and barn to avoid losing your way and freezing to death. News on TV told of a long cylindrical load stranded on an eighteen-wheeler that had been forced to stop in Liberal. Speculation ensued as to what it could be. Maybe it was a Jupiter or Thor, but most likely it was an Atlas on its way to the Cape. National security was at stake. Guards, shivering in the darkness, had surrounded the cargo throughout the long night. I was thankful the blizzard had not yet swept eastward to Wichita.

Ron Gallop came over the next evening carrying something heavy wrapped up in a blanket. "What is it?" I asked.

"Let's go downstairs," he said, looking up from the bundle.

In my downstairs bedroom, he began to slowly unwrap the object, beaming triumphantly. It was a metal tube about two feet long and four or five inches in diameter, shaped like a coke bottle at one end. It looked like it was made of some type of stainless steel and had several pipes and fittings coming out of its closed end.

"What is it?" I repeated.

"It's a rocket engine. A vernier engine for an Atlas."

My mind turned to a vision of Western Kansas, the blizzard, the impossibility of what was before us. "How?"

"My dad's company. I smuggled it out of Production Machine." Ron had explained to me that Production Machine Company had a contract to manufacture these engines.

"How did you do that?" I asked.

"I borrowed a key and made a duplicate."

"That's awesome!" I said. "You could get in serious trouble."

"No problem, I'll have it back in place by tomorrow morning and no one will know the difference."

I grabbed a quadrille pad and began to make a sketch. It was evident that this was a regeneratively cooled engine of a type I had read about. Fuel is circulated around the combustion chamber to keep it cool before being injected into the chamber for burning. The heat absorbed by the fuel is not wasted but adds to its energy. But how much fuel is circulated? What about the oxidizer? What metal is the engine made of? We discussed these things with great excitement during the next half hour, even though we were quite certain the answers to these questions were all top secret.

At last, I ventured, "You know, I think we might be able to build something like this for our next rocket."

* * *

Production Machine Company was located in a nondescript building in the gritty end of North Wichita. Ron had taken me there several times. There was a small office in the front where nobody worked, with a door leading to a large area filled with machines and machinists. Once, we started talking with a worker who was using a jig boring machine to make a precise hole in a large casting he said was part of the erection system for a Nike missile. Maybe we distracted him by talking, but he made the hole a bit too large. Ron's dad had to call someone to get permission to plate back on a few thousands of an inch to bring it back into spec. I felt bad about it.

Four Nike-Hercules sites were being installed near Kansas City to protect against Soviet bombers and missiles. Each rapid response Nike-Hercules carried a 20 kiloton W31 warhead with a range of 75 miles, meant to destroy entire formations of hostile bombers or missiles without requiring a direct hit. The city might be saved, but the surrounding countryside would be obliterated. Production Machine was one of the contractors supplying parts for these defense systems.

Our favorite machine was the big Jones and Lamson turret lathe. A long aluminum cylinder about 3 inches in diameter would be fed into the lathe's chuck at the push of a button, followed by tools on a six-position turret that advanced into the work one by one. Before the big final cut was made, the operator would deftly step back a few paces to avoid being hit by a hot ribbon of aluminum which would shoot into the air and fall to the floor. After the finished part dropped into a bin, the whole spectacle repeated.

Before long, I began to soak up the lore of these machines and the smell of machine oil. I realized if we were ever going to do something big with rocketry, we needed to figure out how we could make all the parts for Cryolite.

* * *

Ron and I had been thinking for several weeks about how we could make nozzles for both the first and second stages of our awesome Cryolite rocket. We began to formulate plans for breaking into Production Machine Company. Ron had long since returned the Atlas vernier engine, but he still had a door key, so we had a way in.

Days before the planned break-in, I had gone to The Yard's stockpile and purchased eight inches from the end of a long solid aluminum cylinder the salesman had cut off for me. I had previously made a cross-sectional drawing of the nozzle. We would hollow out the inside and outside into the shape of a nozzle and later have it welded to Cryolite's tube.

We struck after midnight, bringing along my brother Dave to serve as lookout. I parked my Ford a block away. Sticking to the shadows, we walked along a row of old buildings before arriving at the front door of Production Machine. Perhaps the key wasn't a perfect copy or else it was too dark to make out the keyhole, or maybe he couldn't keep his hand from trembling—Ron kept fumbling with the damn lock for several minutes.

"Can't you hurry it up?" I asked, glancing up and down the narrow street. "Somebody's going to spot us." The key fit, the knob turned, and we were in the front office. The machine room was dark. As our eyes adjusted, we could make out the shapes of the machines illuminated by a few light bulbs in the alley behind the building.

"Dave, I think we can keep an eye on the front. Why don't you stay near the alley." He made his way in the darkness toward the windows.

With our lookout posted, Ron and I moved over to the Logan engine lathe he had pre-selected as our machine of choice. It was a big lathe, plenty big for the job, but neither of us had much of a clue as to how to operate it.

We turned on a small lamp attached to the headstock and proceeded to lock the cylinder in the three-jaw chuck. By 2:00 am, we had used a large drill to bore out the center hole and we were beginning to make passes with the compound set at twenty degrees to form the nozzle's exit cone. We conversed quietly as the chuck spun around with a whirring sound.

"See anything?" I called out to Dave. The distinct and intoxicating odor of roasting potato chips wafted in from the Lay's factory across the alley.

"Just a few guys taking a smoke once in a while," he said.

It began to rain. After a while, we could hear it pounding on the roof. "Good," I said to Ron, "This is perfect. There's no chance anyone can hear us now." By 4:30 it was still raining, and we had finished the interior cone. Things were going well, and we had pretty much held to the tolerances on the drawing. Suddenly, we heard a pounding noise on the wall near the lathe. "Did you hear that?" I said.

We shut down the spindle and let it coast to a stop. Nothing, only the sound of rain. After a few minutes we reversed the nozzle in the chuck to prepare for machining the entry section. We ran up the spindle again and soon another loud knocking was heard. We shut down the spindle.

"What do you think it is?" asked Ron.

"I have no idea, but it seems to be coming from next door," I said. "Do you know what's over there?"

"No idea."

We started the spindle again and the pounding started again, louder this time. We shut off the spindle and turned off the light. A funny feeling began to well up in my throat as we stood there in the dark.

"Anything out back?" I called out to Dave. He had wandered away from the window for a while but returned to take a look. He came back across the room.

"There's some red lights in the alley," he said. We made our way to the windows. Rivulets of rain streamed down the panes making it hard to make out shapes beyond. But it only took a few seconds to realize it was a phalanx of police cars.

"Oh... my God!" I said.

Before long, a terrific pounding rocked the front door. My heart skipped some beats—convulsed with thoughts of going to jail. We made our way toward the office. We knew we had to open the door and so we did. Three policemen rushed in, rain dripping off their yellow slickers parting slightly to reveal holstered guns.

An officer barked in a loud voice, "What are you boys doing in here?"

Ron spoke up, "We're just, uh, making a rocket nozzle..."

"How did you break in?"

"I have a key to the door," said Ron, smiling meekly.

Then, Ron's dad appeared at the door. He looked sternly at his son, then stared at my brother and finally at me. After a long silent pause, "That's all right officers," he said, turning to a policeman. "I'll take care of this."

Everyone but Dave and I went outside. The door was cracked open a bit so we could overhear some pretty strong discussions between Ron and his dad lasting for several minutes. After a while, a chastened Ron came back in as everyone else walked slowly back to their cars.

"It's okay," said Ron, managing a smile. "We can finish the nozzle!" What a relief! Mr. Gallop was generous, but I was sure he would follow up with some good punishment later on. It was a great feeling to go back to the lathe, but this feeling was tinged with some anxiety concerning the knocking sound we had heard.

The rain stopped and a rose-colored light was starting to fill the sky as we finished the last cut, shut down the Logan, and un-chucked our nozzle.

Then there was another knock at the door. I opened the door to a slight, middle-aged man standing there in a gray uniform. A badge was clearly visible on his chest.

He said, "Hello, I am from the American District Telegraph." Sure enough, in the center of the badge was a round medallion, with "American District" on the top and "Telegraph" below. "We're right next door. Would you like to see our operation?"

The three of us went next door where a half-dozen men sat at a long desk with telephones, and other gadgets with red and green lights. The man explained, "Here we monitor all the burglar alarms throughout Wichita. It's a big job." Our eyes marveled at all the equipment.

"Were you monitoring Production Machine Company?" asked Ron.

"No," the man said. "We just heard the noise coming through our wall and pounded on it to see what would happen."

12 Parachutes

T HE SPORT OF SKYDIVING was taking hold around Wichita. Harold Truesdell "Smitty the Jumper" Smith was a sign painter and airshow legend who put on a terrific show. Supposedly, he made his first jump in 1922 at the age of 36 with a parachute he made himself out of silk, using a harness made by a horse-harness maker. Apparently, Smitty made a dozen jumps with this original rigging before he even saw another parachute.

Ron Gallop was learning to fly airplanes and thought it would be fun to jump out of them. Somehow, he was able to scrape up money from his paper route to pay for the lessons. He met Reyn White, a jump instructor, who had five hundred jumps to his name. Ron learned how to pack parachutes and how and when to pull the ripcord.

I thought it would be thrilling to be a skydiver like Ron, but instead, contented myself with flying in the Tri-Pacer, and building and launching rockets. It was great fun watching Ron and Reyn freefall and descend under white canopies out of the clear blue skies. Several of us would chase after the pair, helping with the recoveries. Once, the two of them, wearing only swimsuits, parachuted into El Dorado Lake.

* * *

It was early evening somewhere east of Wichita, and the dive plane was circling high in the sky. Phil Roberts and I were chasing after the jumpers.

"There they go!" I yelled to Phil, as we heard the power being cut, and saw two small bodies departing the plane.

The jumpers fell through the sky, with their parachutes popping open long heart-stopping moments later. Suddenly, the calm wind gusted up quite a bit, to about thirty miles an hour. "It looks like they could be drifting into the power lines!" said Phil, "Let's go!" I hadn't noticed the line of high voltage towers a half-mile away from us. We started off through the pasture toward the power lines, running as fast as we could go through the darkening light.

"Looks like they're going to hit them," yelled Phil, running at a fast pace, when the jumpers seemed only a hundred yards above the towers. What would we do if they wind up hanging from the high voltage lines?

Suddenly, I felt a searing pain across my legs, simultaneously hearing a cry from Phil. The pain was joined by a strong force pushing us backwards, away from the power lines. Our legs were still churning, but we weren't moving forward. In the next instant, I knew we had run headlong into a single strand of barbed wire, stretched across the field a short distance below crotch height. It took another instant to realize that the wire, now cutting deeply into our legs, was electrified.

"Walk backwards!" said Phil. Dazed, and shocked by the high voltage, our only recourse was to walk backwards on the rough ground for twenty feet or so to free ourselves from the barbs. We somehow managed to do it while overcoming the pain and convulsions in our legs.

At last, we were free of the barbed wire. We stood there, examining the rips in our jeans and the bloody puncture wounds on the front of our legs, fortunately located barely below our vital equipment.

We had almost forgotten about the jumpers, but when we looked up, they were landing on the near side of the power lines. They were okay, and a car was coming to pick them up. Phil and I, cursing furiously, limped back to my car, and headed home to dress our wounds.

* * *

While Ron was jumping out of planes, we worked on the parachutes for both stages of Cryolite. The design was based on some material I had found in the library a year earlier, drawn up at the end of May 1960. Twelve gores, each spanning fifteen degrees, were made using parachute nylon purchased from The Yard. The rigging lines were #18 nylon cord, also from The Yard, running across the canopy top's five-inch diameter vent hole under nylon jackets. All edging and seams were to be sewn using #40 cotton thread and cotton twill tape. Audine lovingly assembled the five-foot parachute, using her mother's sewing machine.

We needed a way to test it. We decided to cast a lead weight, approximating the weight of the Cryolite stages, attach it to a parachute, and throw it out of an airplane. Gallop and I collected a number of lead weights lying about in the streets, cast off from spinning car wheels. To this collection, we added some lead weights from his dad's fishing gear.

There was a gasoline-fueled blowtorch in Ron's basement. "It's empty, but there's a can of gas in the garage." He ran upstairs and came back with the gas. After unscrewing the fill plug, he instructed "Hold it in position," as he poured the gas. "I've seen my dad do this," he said, screwing the pump in, and pouring some gas in the little tray underneath the burner. An intoxicating smell rose up as Ron touched a lit match to the tray of gasoline, which flared up around the blowtorch barrel.

After a few minutes, the burning gas in the tray had died down, and Ron pressurized the gas tank by furiously pumping on the plunger. We put the lead weights in an empty juice can, and with a twist of a valve knob, the blowtorch roared into life with an intense blue flame and the sound of a jet engine.

Soon, with the blowtorch playing on the can sides, the lead began to melt, emitting noxious yellowish fumes. "Let's try not to breathe this crap," I said, as we jerked our heads back. We kept feeding more lead into the mixture.

After a few minutes, while occasionally pumping on the blowtorch's gasoline tank to keep it pressurized, we had a can full of liquid lead. Using tongs, we stuck a large screw-eye into the pool of lead, shut off the blowtorch, and held the screw-eye in place until the lead solidified. Finally, we dropped the can into a bucket of water where it let off a satisfying hiss and cloud of steam.

With the juice can peeled off our lead casting, we had a weight of about five lbs. Audine had finished the five-foot chute, so we attached it to the weight and deployed it while running into the wind. It always opened with a nice popping sound. But we needed to test the parachute deployment at high speed and at altitude. It was time to confront my father with the question of throwing the parachute out of our Tri-Pacer.

"Do you understand what would happen if the parachute hit the tail? We could go out of control and crash," he said.

"I could lean out the door and throw it down so it wouldn't open until it was past the tail." After much discussion, Dad finally acquiesced. Ron would drive to the farm, acting as observer, while Dad and I flew over it in the Tri-Pacer.

* * *

We lifted off from Ken-Mar in the late morning of a cloudy day, leaving the pattern and heading southeast toward the Hollis farm. I had wrapped the parachute's lines around the lead weight, and packed the canopy as tightly as I could, emulating how it would be ensconced in the Cryolite's second stage.

We had allowed time for Gallop to drive to the farm, and before long we spotted him in the field below, holding up a ribbon showing the direction of a light breeze. Dad nosed into the wind at 2,000 feet above the ground and slowed to 100 miles per hour. I snugged up my seat belt, and on his command, twisting my body to the right, forced the door open with my left arm against the force of the wind. I threw the parachute downward as hard as I could with my right arm, taking care to miss the wing struts. Looking back, I saw that it had missed the tail.

The chute opened successfully and drifted to the ground where Ron was able to recover it and return with it to Wichita. Upon inspection, all of Audine's careful stitching had held, and the parachute was undamaged. We had one more important part of our Cryolite rocket! She soon completed the six-foot parachute for the first stage, but it didn't seem necessary to test it separately.

13 Computers

I BEGAN TO SPEND night and day at the University, continuing my work on the 610, while my math classes suffered. I became such an expert in its operation that some faculty and graduate students regularly consulted me on programs they were writing. One professor needed to do some physics calculations, so he handed me a page of equations to run on the computer. I had no idea what the math was all about, but I could do the programming and he paid me for the results!

One day, to my dismay, the IBM people came in and took away the 610! Unbeknownst to me, it was rented, and the University had decided to stop paying for it. I still had many calculations to perform, even though I had entered more hours in the logbook than all the faculty and other students combined.

* * *

Shortly after the devastating demise of the IBM 610, another computer appeared in a basement room of the engineering building. The light blue Bendix G-15 was the size of a large refrigerator. Immediately, I began to learn how to program it, immersing myself for days in the 214-page manual where I read about a growing need for computers to solve two types of problems.

The first are, *"problems whose solutions are essentially simple in nature but which must be solved over and over again, each time for a different set of values."*

The second are, *"problems whose solutions are so complicated that men cannot spare the time and effort to solve them by the pencil-and-paper method."*

My rocket calculations were of the first type—for example, calculating the motor thrust for many different values of the pressure and temperature inside the combustion chamber. Years later, working in California, I would be confronted with a problem of the second type.

The G-15 had vacuum tubes, relays, and a rotating magnetic drum for a memory, much like the 610. It also had paper tape and a typewriter for the input and output. But unlike the 610 which was programmed in the decimal system, the G-15 used the binary number system where values are represented by a string of 29 ones and zeros, called a "word."

As a shorthand, the sixteen-value "hexadecimal" number system was introduced, with digits 0-9, and u, v, w, x, y, and z. Thus, the decimal value 123,456 is expressed in "hex" as 1y240. I learned that the G-15 was an example of the "von Neumann" architecture, where both data and commands could be stored and manipulated in memory. The IBM 610 only stored commands on paper tape.

In the G-15, there was no difference in appearance between data numbers and commands. They were only distinguished by the time at which they are accessed on the rotating magnetic drum. After many unsuccessful days of trying to understand how to program this complicated machine, I realized it would require changing all my previous work.

The manual was so detailed and difficult I despaired of ever understanding it. Strangely, I read a bit of levity half-way through: *"The situation arises when someone other than the programmer, himself, attempts to twist the program to suit his own needs, heedless of warning. In this case, the programmer might very well like the output to consist of a few well-chosen four-letter words..."*

At another point I read: *"The simple 'decision-making' power of computers is what has led laymen to use the term 'electronic brain,' and other equally erroneous terms, when referring to computers. You can see that, actually, the G-15 does not 'think;' it merely tests, upon command, the condition of a circuit or component as to 'on' or 'off'."* I had first learned of the fascinating term "electronic brain" in grade school while reading *My Weekly Reader* and had spent many hours lying awake at night trying to imagine how it could work. I recalled that an "electronic brain" at MIT had been used to track the orbit of Sputnik.

One day, the Bendix salesman dropped by, and asked me how it was going. I expressed some frustration about my lack of understanding of the G-15 manual. He told me that almost no one programs the G-15 in the so-called "machine language," covered by the manual. He loaded the company's newest program into the computer, a language called "Intercom 500."

With Intercom 500, one could write programs using a sequence of commands input as decimal numbers followed by the decimal number of the data location. For example, simple addition was performed by the command "43ADDR," which added the contents of ADDR (expressed as a decimal location in memory, such as, 1432) to the contents of a special location called the "accumulator," placing the sum in the accumulator, and leaving the data in ADDR unchanged.

Commands from 40 through 49 performed arithmetic operations; 20 through 29 transferred control; 30 through 39 governed output operations; 50 through 55 were used to input commands and data into memory; and 61 through 69 started and stopped the computer. Command 63 rang the bell, which I found handy for timing programmed loops.

After several intensive weeks of study, while barely managing to stay afloat in my Differential Equations class, I felt I had, for the most part, mastered programming the G-15 using Intercom 500.

* * *

I was filled with excitement over the news that the Echo satellite would pass over Wichita. I had never observed a satellite, so I decided to get up at 4:00 am to view it passing overhead. Echo was a 100-foot diameter aluminized Mylar balloon. Radio and TV signals could be beamed to it and reflected off its surface to reach stations hundreds of miles away. The night before, I tried to get my parents to get up with me. Dad wasn't interested and declined, but Mom agreed.

It was clear and chilly as we stood together in the backyard searching the dark sky. ``Do you think you might go into astronomy?" she asked.

"Probably not. I'm not really sure what I'll wind up doing. Maybe something in computing if that becomes a profession."

"Your father became an architect at a time when people were saying there was no need for architects because there was no money to build anything." Mom and Dad courted during the Great Depression and were married in 1937. Now nearing fifty, she had been through a lot.

She was born on an impoverished farm near Denmark, Kansas, northwest of Salina. Her parents were peasant farmers who had emigrated from Sweden to escape disastrous crop failures. Her father died when Mom was only three months old. The house they lived in was little more than a shack with broken windows that let the snow in.

She had told me she suffered from several serious bouts of pneumonia as a young child. Her mother was unable to continue with the farm and was forced to hand Mom over to relatives. At age 10, she sold doilies and other items from door to door and worked ironing clothes to help pay for her upkeep. Her older sister contracted scarlet fever and died at age sixteen. Mom had an interest in art, graduating from high school, but she never attended college. Maybe I inherited her love of art.

"I'm getting cold. Do you think the satellite will show up?" she asked.

"Probably any minute now. Maybe the newspaper had the time a bit off." As a small child I remember lying in her warm lap listening to the sound of her voice reading stories. "There it is! Look over there!"

We were thrilled as Echo moved slowly through the sky, making its way nearly overhead against the faint stars in the early morning. Mom wanted me to succeed in my science dreams; Dad, I believe, would have rathered I follow him into architecture.

* * *

I had read in a magazine that the solid-fueled Minuteman missile under development had cast propellant with a star-shaped hole down the center of its combustion chamber. The star shape provided increased burning area over a simple circular hole, and if designed correctly would give a nearly constant thrust level. That gave me the idea to design the ideal star shaped central hole for the Cryolite stages. It meant solving the challenging inverse problem: given a desired thrust versus time profile, I had to generate the star geometry that would produce it.

I started to work on the problem, but its complexity was quickly overwhelming. Moreover, I began to realize that the burning gases moving from the combustion chamber toward the nozzle would rapidly gain speed. This would cause additional erosion of the propellant charge, another factor that would need to be taken into account. I spent many days and nights working on the problem with the help of the G-15.

One day, shortly after noon, the Bendix salesman showed up again, this time accompanied by two workmen. "We're here to take away the G-15."

"What?" I felt heartbroken. I had spent so much time and energy learning how to use the G-15 that I felt cheated. "Why do you need to take it away?"

"The computer is just here for a trial period. The University decided not to buy it."

I pleaded with him. "Can you give me a few more days to complete my calculations?"

"Sorry. The company needs to take it back today." He pulled the plug, and I sat there listening to the magnetic drum winding down. One workman pushed the front edge of a hand truck under its side, while the other cinched a strap around its middle. So that was it. I watched as the G-15 was hauled out, taken down the hallway, and into the elevator. When the elevator doors closed, I felt as if I had lost a dear friend.

The salesman saw how disappointed I was and hung around for a few minutes. He asked me about my calculations and how they were being executed on the G-15. As best as I could, I told him about my rocket calculations and the internal ballistics of gas flow. Finally, I explained how I was calculating the star shape of the central hole.

Then, he seemed to take some pity on me as he said, "I will put you in touch with some engineers at Beech Aircraft Company. They have two G-15s. Maybe you can use them to complete your work."

* * *

Before long, I had made arrangements to visit Beech Aircraft, only four miles from home. There, I was given a special entry badge letting me in to the computer center most any night at 12:00 am.

The day after I got my badge, I could hardly wait for midnight to arrive. I had gathered up my notebooks and programs punched on paper tape. The guard waved me in. I was greeted by two programmers, who showed me the side-by-side pair of light blue G-15s. Nearby was an older Burroughs E-101 computer, which could be programmed only by wiring a patch panel. One of the G-15s was busy in use, working on some airplane calculations. As I loaded my program on the other G-15, I felt as if I had died and gone to heaven.

I made a lot of progress that night, but I knew I would be coming back a lot. A G-15 was not available every night; I had to call in before midnight to ask if one was available. I continued to work for several months, averaging around three midnight visits a week, usually working until 2:00 or 3:00 in the morning. The Beech programmers seemed interested in my work, and it was a pleasure visiting with them as the G-15 chugged along.

I was making good progress—but it meant it was hard for me to stay awake in my morning Complex Variables class. Eventually, I found a satisfactory star-shaped hole, and I carefully plotted it on polar graph paper from the numerical results printed out on the G-15's automatic typewriter.

On several occasions, both G-15s were available, so I ran trajectory calculations on one, while running the gas flow calculations on the other. I found that the Cryolite rocket would attain a higher peak altitude if a delay was introduced between burnout of the first stage and ignition of the second stage. This was true because of the added momentum from the burned-out first stage and because the second stage would operate at higher altitude in thinner air where drag would be reduced. I was able to calculate the optimal delay time, assuming the two stages remained attached until second stage ignition.

* * *

By late May 1961, we were making excellent progress building the Cryolite rocket stages. We had completed both parachutes and tested one of them. We were busy working on a method for attaching fins to the body. After contacting some east coast companies about propellants on our Propulsion Research letterhead, we purchased and soon received a 100 lb. drum of zinc dust powder and a 50 lb. box of sulfur. We were trying to figure out a safe method for making the castings.

Calculations on the G-15 showed Cryolite's second stage might reach a minimum of 35,000 feet of altitude at its apogee. We felt this could be a record for an amateur-built rocket. At over six miles high, its flight was too risky to launch from either the farm or the ranch. We didn't want to accidentally shoot down an airliner or B-52. Instead, we had received an invitation from the U.S. Army Artillery and Missile School at Fort Sill, Oklahoma—250 miles southwest of Wichita. We could launch Cryolite there and they would track it by radar. Several years of hard work were about to pay off. My excitement was palpable.

* * *

Audine and I had been with each other a lot, going on drives in the country, to the movies, and having fun with each other's families. We had been going steady for over two years since our first kiss under the streetlight at the corner of Oliver and East Morris. We were in love, but until recently had resisted sexual intercourse.

Audine was a month away from graduation when she met with me to tell me some startling news. "I missed my period this month. I think I might be pregnant."

"How can that be? I thought we were being so careful."

A week later, after a visit to the nearby clinic, she told me she was indeed pregnant. I was alarmed, panicky, and felt sick. It didn't take long for me to realize our lives had changed forever, and this was the end of my rocket dreams.

Figure 1: Ralph Hollis, MOUSE model and Vulcan I rocket.

Figure 2: MOUSE model in Ralph's home lab.

Figure 3: Ralph Hollis and teacher Joe Foraker,
Wichita Eagle photo, June, 1956.

Figure 4: Ralph and Vulcan II rocket.

Figure 5: L-R: David Hollis, unknown, unknown, Harold Wiebe.

Figure 6: Rocket Boys L-R: Steve Hawkins, Harold Wiebe, Phil Roberts, Ralph Hollis.

Figure 7: Behind a barricade at the Hollis farm, L-R: Phil Roberts, Ralph Hollis, Wendell Allen, Bruce Bullock, Steve Hawkins.

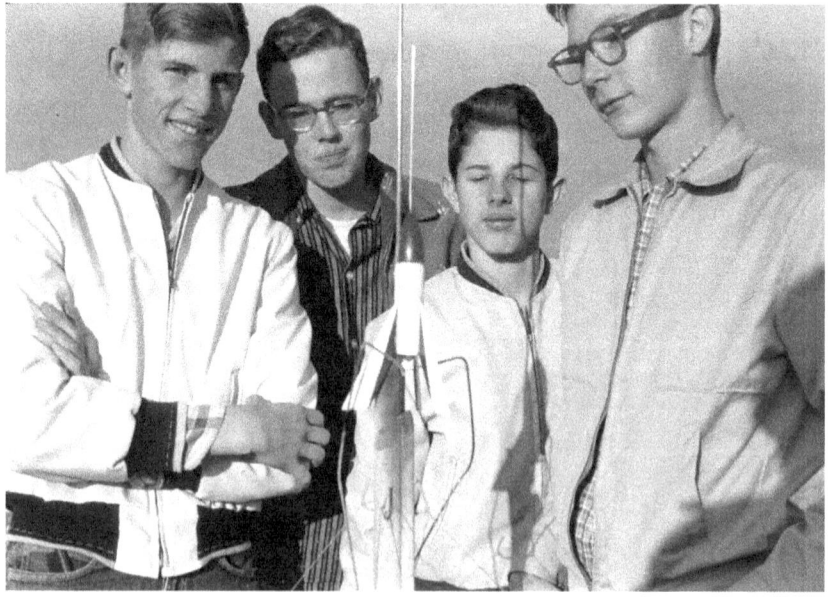

Figure 8: Firestreak rocket on launch stand. L-R: Ralph Hollis, Steve Hawkins, David Hollis, Gary Peyton.

Figure 9: Firestreak rocket launch.

Figure 10: David Hollis, manning tracking station.

Figure 11: Static test stand at the Hollis ranch, testing Arcturus motor.

Figure 12: Ron Gallop and Jon Freeman ready for Arcturus motor test.

Figure 13: Ralph Hollis, Southeast High School senior, with prototype of Hyper-Force rocket.

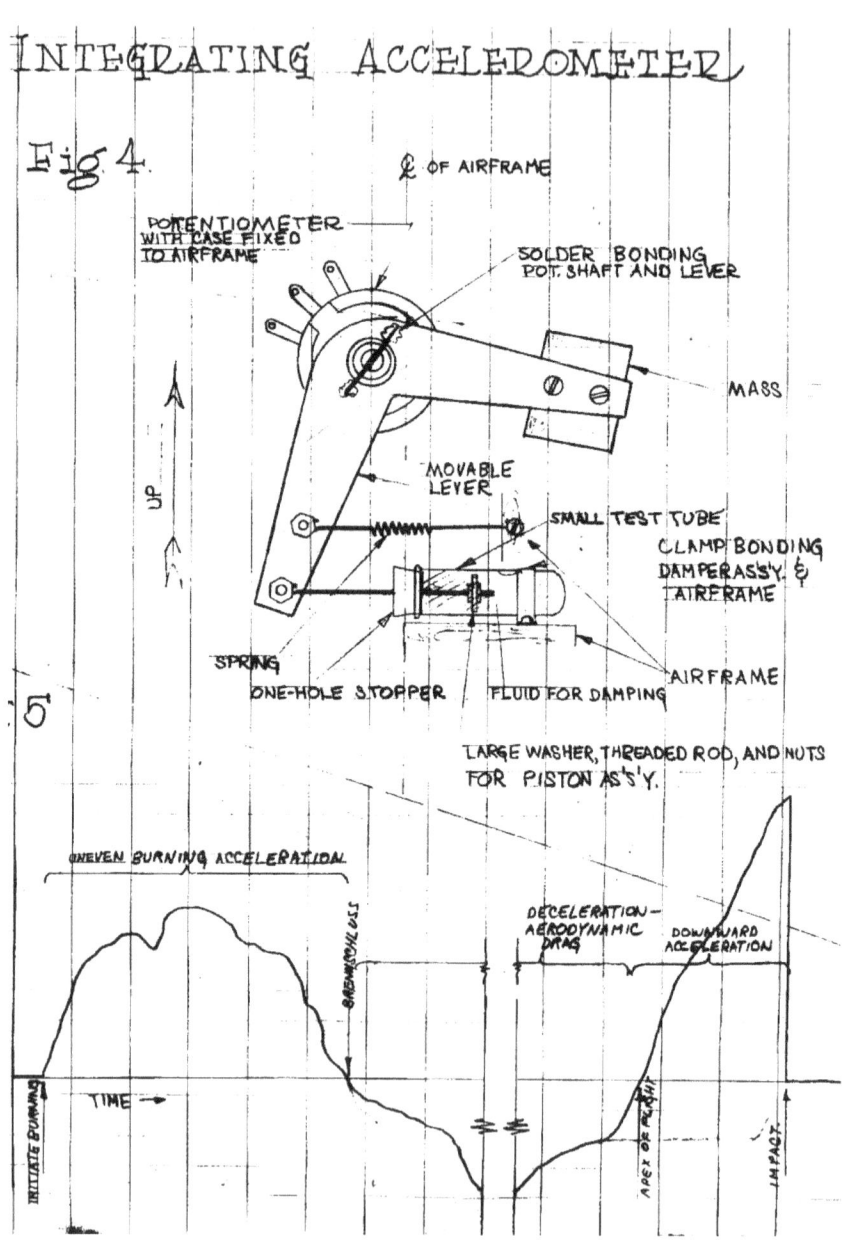

Figure 14: Harold Wiebe's 1957 design for an integrating accelerometer, part of his transistorized Parachute Timing Release Computer (PTRC).

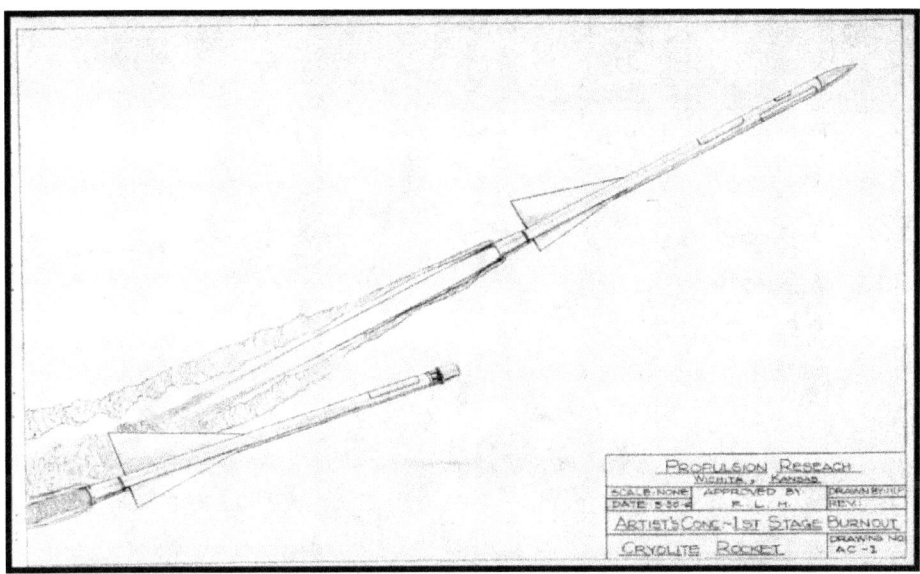

Figure 15: Visualization of Cryolite 2-stage rocket at first stage burnout,
Drawn May 30, 1960.

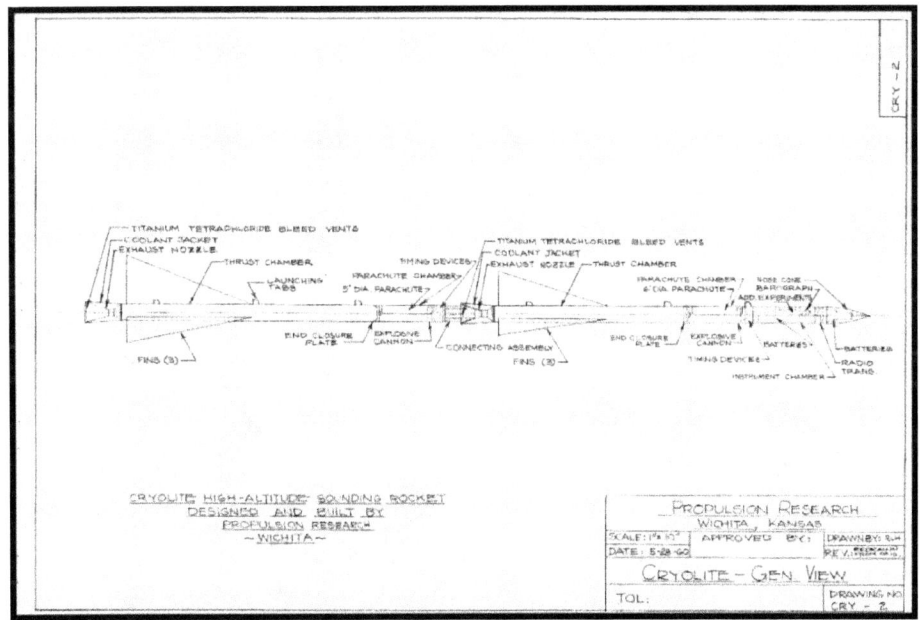

Figure 16: Details of Cryolite 2-stage rocket, May 28, 1960.

Figure 17: L: Reyn White, R: Ron Gallop.

Figure 18: Marc (Rocky) Roth and Ralph Hollis in Professor Dragsdorf's crystal structure laboratory at Kansas State University.

Figure 19: Autonetics building 235.

Figure 20: Autonetics site at the peak workforce of 36,000.

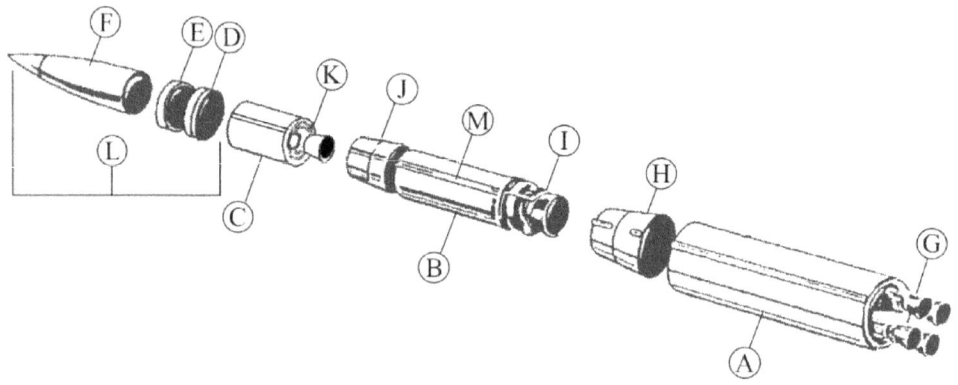

Figure 21: Minuteman III: (A) First stage; (B) Second stage; (C) Third stage; (D) PBPS, Post-Boost Propulsion System; (E) PBCS, Post-Boost Control System; (F) Shroud covering RVs, decoys, and chaff systems; (G) Steerable nozzles, protected by cylindrical skirt, omitted for clarity; (H) Interstage skirt; (I) Second stage LITVC, Liquid Injectant Thrust Vector Control; (J) Inter-stage skirt; (K) Third Stage LITVC, Liquid Injectant Thrust Vector Control; (L) PBV, Post-Boost Vehicle assembly; (M) Cable raceway (all stages).

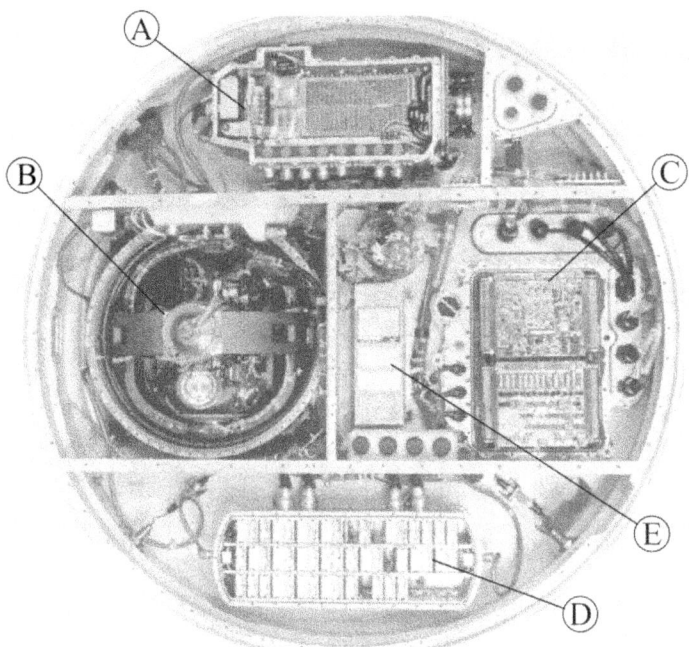

Figure 22: Post Boost Control System (PBCS), covers removed: (A) D37D computer (the *brain*); (B) Inertial Measuring Unit (IMU) (the *heart*); (C) IMU electronics; (D) amplifier; (E) battery.

Figure 23: Autonetics Skywriter calendar girl Carole Murillo, secretary in SAS Sub-Contractor Management, and year-around sun worshipper.

BOOST VEHICLE DYNAMICS

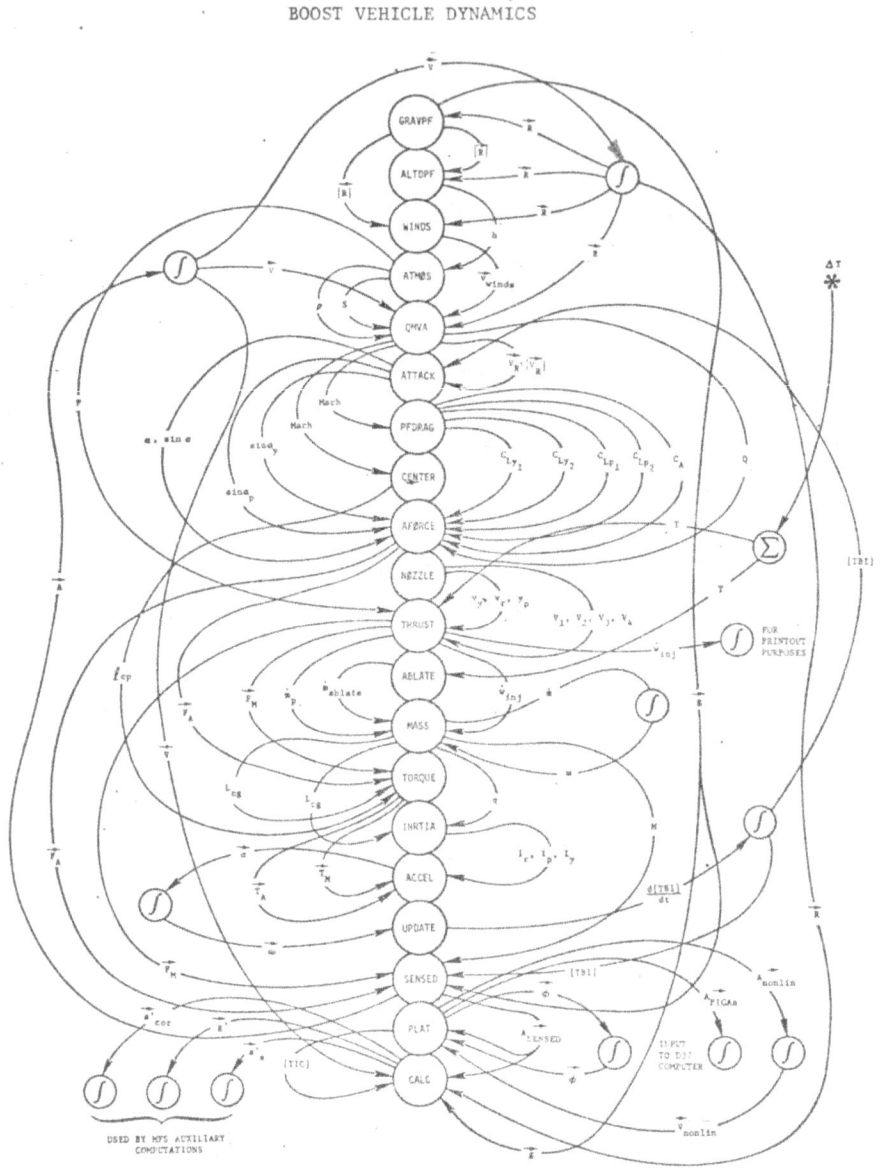

Figure 24: MFS Boost Vehicle Dynamics: Center circles represent routines called in sequence from top to bottom in a single Runga-Kutta pass. Curved arrows indicate information flow between routines with integration and summation operations indicated by small circles.

Figure 25: L: Photo of FTM 201 launch; R: Autonetics Minuteman III first launch full newspaper page announcement.

Figure 26: Minuteman III RVs and shroud. (Photo: Department of Defense.)

Figure 27: Photo of dummy RVs impacting near Kwajalein Atoll (DoD file photo).

Hollis

Figure 28: Photo of Ralph Hollis in Autonetics Skywriter newspaper (July 25, 1969).

Figure 29: Carol, Ralph, and Dana in Wichita.

Figure 30: Greg and Marsha wedding.

Figure 31: Larry Hambly and Ralph's MGB.

II
KANSAS STATE

14 Rocketry Interrupted

IN LATE SPRING, the trees were all dressed in bright shades of green. Audine and I drove to College Hill Park in the early morning where we found an isolated bench where we could sit and talk. Looking directly at me, she said, "What should we do?"

I looked into her eyes. "Do we love each other?"

"Yes," said Audine.

A small rabbit bounced along near the edge of the street. "Mom and Dad think we should get married. I think we should."

"Yes," she said. "Me too. Will you help care for the baby?"

"I'll try to do my best." The words were coming out, but I had no idea what they meant. *How can I possibly be a twenty-year-old father?*

The night before the wedding I was in bed trying to sleep. Brother Dave was in his bed on the other side of the room. Mother entered silently, leaned down, and kissed my forehead. Overcome with emotion, I fought back tears. When she left the room I began to sob uncontrollably, hoping my brother didn't hear me. I couldn't stop crying until much later when I eventually fell asleep.

The small family wedding was in June—bride in a beautiful white gown, groom in a black suit and tie. Brother Dave was Best Man; Audine's sister Marjorie was Maid of Honor.

After the wedding, Dad and I drove to Kansas State University in Manhattan, 130 miles northeast of Wichita. He had convinced me I would be better off going to school there, at his alma mater. Also, a housing development was provided on campus for married students. We met with A. B. Cardwell, head of the physics department. Dad knew Professor Cardwell, having previously done some design work for him. After telling him I had been a math major in Wichita, and describing my interests, he persuasively argued I should switch to physics at K-State. Since I had not done well in my math classes, my long-term goal was to be a physicist anyway, so I agreed. Starting in the fall, I would be a physics major.

Back in Wichita, instead of working at Reiss and Goodness Engineers again that summer, I spent the months driving a tractor in the fields near Douglass, plowing and harrowing the ground for local farmers, earning what I could to support my future family. It was tough work, but Audine and Marjorie brought picnic dinners that we ate together in the early evening light. By end of summer my skin had turned a dark brown.

In late summer, 1961, we moved into an apartment in Jardine Terrace, walking distance from Willard Hall which shared the physics and chemistry departments. I was assigned a desk with several graduate students in the corner of a physics lab. Audine was happy at the prospect of motherhood. She seemed content to set up housekeeping, buying our groceries from a small store near campus.

That fall, I felt disoriented; worried about our future with a baby coming soon. But for both of us our marriage was a great new adventure. Things were working out, thanks to our determination and financial help from my parents. As a junior physics major, I took Calculus, Mechanics, and miscellaneous required classes in liberal arts. Dave was now a freshman at Kansas State, planning a career in architecture, living in the Goodnow Hall dorm.

I struggled in Mechanics, unable to keep up with what Professor Ellsworth was saying. Freshman physics hadn't been required for math majors at the University of Wichita; switching from math to junior-level physics was too challenging for me. The bad news came in December: I received an F in Mechanics. I was sorry for my parents' misplaced faith in me and for letting down Audine, and perhaps most importantly—I lost confidence in ever becoming a physicist.

Fall transitioned to a snowy January. Audine's due date was coming up fast and I was having trouble starting my Ford in the cold weather. For a week, I set the alarm at 2:00 am to get up and start the car, letting it warm up for a while. One early morning, she said, "It's time." My heart raced as I turned the key. The engine turned over. It started! We made it to the hospital in time. Later in the morning, a new baby girl entered the world. I suggested we name her Carol

Anne, because I thought it sounded pretty. Audine agreed. Our lives had changed again.

Audine learned to breast feed. That spring I got reasonably good at burping and changing diapers, marveling at Carol's soft skin. I learned to be careful to avoid sticking her with a pin. Baby Carol needed a lot—changing table, pacifiers, diapers and burp cloths, wipes, baby bathtub, baby shampoo and powder, thermometer, and who knows how many other things. We didn't have much money, but somehow we survived. Mom took the bus up from Wichita to help out, a big blessing. I enjoyed our happy little family.

Chastened by my failure in Mechanics, I put in a terrific effort for my E&M (Electricity and Magnetism) class. At semester's end, to my great relief, I had earned a solid B. My lowest semester grade was a C in Psychology. The summer was spent, in part, going over the set of problems in Mechanics—trying to understand why I had done so badly. Professor Ellsworth would allow me to repeat the course in the fall. It was beginning to dawn on me that perhaps doing well in my schoolwork would first require laying aside my thoughts of rocketry.

* * *

I met Dave in the commons area of the Goodnow Hall dorm on October 22, 1962. Dozens of his classmates were gathered to hear President Kennedy's address to the nation. At 6:00 pm, the President announced to a stunned and sober group of us that the Soviet Union had placed nuclear missiles in Cuba.

"Good evening my fellow citizens, this government has promised, has maintained, the closest surveillance of the Soviet military buildup on the island of Cuba. Within the past week, unmistakable evidence has established the fact that a series of offensive missile sites is now in preparation on that imprisoned island. The purpose of these bases can be none other than to provide a nuclear strike capability against the western hemisphere..."

We all got ready for the inevitable World War III, until conditions eased up five days later after Kennedy's successful naval blockade and agreement with Soviet premier Khrushchev to withdraw our U.S. short-range nuclear missiles from Turkey and Italy. The world had come a hair's breadth away from annihilation.

* * *

When I took Mechanics for the second time, I pulled a solid A. Professor Ellsworth was as surprised as I was. At semester's end we packed up and drove home to share Christmas with our families who had a good time fussing over the baby.

Dad and I took a drive out to the farm. At the Douglass train station, we watched as enormous arc-shaped steel rebars were being unloaded by crane from a line of flatbed railcars. "They must have a six-*eench* diameter cross section." Dad had a peculiar way of saying *inch*.

"I've never seen a rebar that thick," I said.

"You haven't seen anything yet," he said. We drove seven miles east to a huge construction site, as much as three acres, excavated to a depth of forty feet. At one end of this pit, we saw an enormous hole in the ground. "It's more than seventy feet across and deep enough to bury a fifteen-story building." Dozens of cranes and hundreds of workers down in the hole were placing the rebars. "It's for a Titan missile," said Dad. "Someone told me they are making the silo walls six feet thick so it can survive a nuclear attack."

The Titan II missile complex, number 533-6, was one and a half miles from our ranch where we had launched Arcturus only a few years before. Another silo, number 533-7, was going in near Rock, Kansas, three miles south of our farm where we had launched the Firestreak rockets. These sites were two of eighteen installed along an arc south and east of Wichita. Each site would house an enormous liquid fueled rocket with a nine-megaton warhead. Their fuel was unsymmetrical dimethyl hydrazine, and the oxidizer was dinitrogen tetroxide. Each missile had over 315,000 lbs. of this highly toxic and volatile propellant that would explode on contact.

The Titan sites were surely pinpointed as prime targets in the crosshairs of Soviet ICBMs. The Cold War had moved ever closer to me and my family.

* * *

My senior year in physics included courses in quantum mechanics. I used an analog computer in the Department of Engineering to solve the Schrödinger equation. There was also an IBM 650 computer I learned to program. It had two-thousand decimal digits of drum memory and used punched cards for input and output. I particularly enjoyed my work in a lab course where I performed the Millikan oil drop experiment, measuring the charge on a single electron. I was intrigued that an electron has an electrical charge and a mass, but as far as we know has no physical size. For the first time, I began to understand and feel the beauty of physics and its study of matter, fundamental particles, energy and forces, and how motion evolves through space and time. I was hooked!

My old friend Phil Roberts, now at Caltech in Pasadena, was designing a mobile robot, an improved version of our wall-following robot. This one was to employ two motors driving the wheels through a pair of differentials. He needed some parts machined so I made them in the physics shop out of scrap aluminum and sent them to Pasadena.

By semester's end, only six or seven students remained in my physics cohort. All were smarter than me, garnering scholarships to various graduate schools. I graduated at the bottom of the class with a grade point average of 3.1, a solid and respectable B.

* * *

That fall, I started work on my master's degree in physics at Kansas State. A brand-new building, Cardwell Hall, had been built. As a teaching assistant, I could at last earn some money. Professor Dean Dragsdorf took me in, giving me a place in his X-ray crystallography lab. There, I met graduate students Rocky Roth, Glen Reese, and technician Richard Hunt.

Rocky and I hit it off right away. His real name was Marcian, which sounded too girlish; the nickname was from Rocky Marciano, the boxer. He hailed from out west, somewhere near Hays, Kansas. Glen was a rugged cowboy type, an excellent guitar player who sang a repertoire of western music. He could sing a terrific rendition of the classic *Mule Train*. His girlfriend was Lee Ann, a dark-haired beauty.

Two other students, Don Burton and Phil Rinard, were a year or two ahead of me studying theoretical physics. I looked up to them as mentors. Often, they could explain the solution to a physics problem too difficult for me to understand, but easy for them. My interest in the physics of flying machines continued, particularly concerning the Frisbee and the Australian boomerang. I made a boomerang out of laminated walnut in the new physics shop and became quite adept at throwing it.

Phil had a new camera he was testing. We fastened small red and green Christmas lights to the boomerang wing tips, powered by a small battery. Phil took time exposures of the boomerang flights to determine forward flight velocity and rotational speed. He then worked out an equation combining aerodynamic lift and gyroscopic precession to completely describe the boomerang's return-to-sender behavior. We started to write this up in a paper when we read that someone had already done it and published it in Scientific American!

Audine went to the doctor in April, finding out she was pregnant once again. This was devastating news to me. Carol was almost two and I was barely able to put enough food on the table for the three of us. When I bought a $20 humidifier for the baby's room the check had bounced. We had heard that a pill had been invented that could prevent conception, but doctors were unsure of its safety. This form of birth control would not become generally available until 1965.

Night after night, lying in bed, we heard the worrisome boom-boom and saw flashes of light from heavy artillery coming from Ft. Riley, sixteen miles southwest of us. Preparations were beginning for a war in a small Asian country halfway around the world that almost no one had ever heard of.

In November, Audine and I were looking forward to a visit from our friend Steve, one of the former Propulsion Research rocket boys, who would drive up from Wichita. Glen and Rocky were in the lab when I came in. Rocky looked at me with a sad face when he told me, "President Kennedy has been shot and killed." I was in disbelief. A portable radio next to the coffee maker was telling us all about it. Steve had heard it on his car radio by the time he arrived later in the afternoon.

I thought about how this terrible tragedy might affect the President's ongoing project to land men on the moon. I recalled his brilliant speech at Rice University a year before. *"We choose to go to the moon. We choose to go to the moon in this decade and do the other things, not because they are easy, but because they are hard."*

My physics courses had problem sets due every week or so. They were hard, too. Sometimes I worked on them alone, and sometimes a group of us students would work together. I scribbled the solutions on scraps of paper that were hardly legible, working into the night. Audine would spend hours transcribing my writing, making it nice and neat on new sheets of paper that I would turn in to the professor in the morning.

Another baby girl was born on the first of December. Audine chose the name Dana Louise. She was a beautiful baby, but Audine had had some difficulty with the birth, and had to stay a few extra days in the hospital. I needed to study like hell for my finals while caring for Carol Anne, so I didn't get to spend much time with Audine and the new baby. Thankfully, Mom was again with us to help out. "You need to spend more time with your wife," counseled my friend Don Burton. I guess I hoped Audine would just "suck it up." Later, I would regret harboring these feelings.

Audine's little sister Carol joined us in the summer of 1964. At fifteen years old, she was a big help taking care of the babies. By this time, we had moved into a two-bedroom apartment; still a bit cramped, but better for our growing family. Audine bought some cloth and she and Carol sewed pretty dresses for the girls. They were proud of how nice they looked. She hung them up on the clothesline to dry one day and someone stole them. All their expense and effort were gone.

I had started work on my M.S. degree, studying the crystallography of histamine and sodium hyponitrite. The latter compound was of interest by the

Air Force as a possible oxidizer for rockets. As a research assistant on the grant, I had a small but welcome source of income.

In late fall, a bad-news telephone call came from Mother. Twenty-two-year-old parachute instructor Reyn White, whose actual name was Stephen Reynolds White, had plunged to his death when both his main and reserve chutes failed to open. He was testing a new type of parachute when something went terribly wrong. I didn't know Reyn well but had watched him and Ron Gallop jump together many times. His death in such a ghastly manner chilled me. Gallop had given up the sport a year before with a total of 50 jumps.

* * *

On Saturday morning, January 16, 1965, at 9:26 am, in Wichita, 130 miles southwest of us, a Boeing KC-135 jet tanker loaded with nearly 31,000 gallons of fuel started its takeoff roll on McConnell Air Force Base runway 36L. Its mission, at the behest of the Boeing Company, was to test the refueling of a B-52 bomber flying somewhere over the Midwest.

A minute later, Ron Gallop, former rocket boy, was driving north past Third Street on Pershing to visit his girlfriend. The giant KC-135 roared over his car at an unusually low altitude as the plane began to cross over Oliver Street. Suddenly, he noticed the air was filled with the strong odor of kerosene.

At 9:30, seventeen blocks north of Ron's location, Audine's older sister Jean Taliaferro Russell was in the living room of her 21st Street apartment tending to her six-month-old, Natasha. She heard a thunderous roar as the KC-135 made a left turn over the University of Wichita, dumping its load of JP-4 fuel. Jean sensed the plane was flying way too low and off a normal flight path. She rushed outside with the baby, scared they might be in the plane's path, crying, "No, no, no," and "Fly, fly, fly!" The KC-135 passed overhead, and seconds later—in a near vertical dive—hit a vacant lot on nearby Piatt Street, exploding in a massive fireball—digging a crater fifteen feet deep. Seven crew members and twenty-three people on the ground perished; the worst military airplane accident in Kansas history.

Audine and I heard about the crash on the radio. "Doesn't Jean live near there?" I asked.

"Yes, on 21st Street. They're saying the crash site is near Piatt and 20th."

We immediately phoned Jean. "The plane passed right over us, splashing the apartment with jet fuel. We were scared but we're all right," said Jean. I felt sorry for the people who died, but relieved that Jean wasn't one of them.

The investigation concluded a malfunctioning autopilot had commanded the plane's rudder to its leftmost limit despite the efforts of the pilot to keep the plane under control. The plane yawed leftward and rolled over moments before impact. The pilots had declared a "Mayday," and were attempting to return to either McConnell or the Wichita municipal airport. Some twenty-nine years later, a similar situation would bring down a Boeing 737 airliner near Pittsburgh, killing all 132 people on board.

* * *

I was doing okay in my classes and excelling in research work. I used X-ray data obtained by one of Professor Dragsdorf's former students to tease out the crystal structure of sodium hyponitrite. This involved writing a huge FORTRAN II program—my first "high level" programming language—for the new IBM 1410 computer. It was a relief to no longer deal with arcane instruction sets.

My crystal structure calculations involved many intermediate steps written on magnetic tapes, to be read on subsequent steps. Glen was a big help in understanding the calculations. Final results were printed out as maps on the IBM 1403 line printer. I then drew contour lines by hand around the peaks in the computed electron density function to locate the positions of atoms in the crystal, a procedure similar to what I had done at Reiss and Goodness Engineers for laying out sewer systems.

* * *

I took some time to cast my Bendix G-15 rocket trajectory programs into FORTRAN II. When I ran these on the 1410 they confirmed my original results of a 35,000 ft. peak altitude for the two-stage Cryolite rocket. Even though I hadn't time nor resources to work on Cryolite, I still couldn't stop thinking of it.

While I was hitting my stride in research, Audine was not doing well, stuck at home with kids, missing out on school and a possible career. We had barely enough money to scrape by. I was spending a lot of time in the lab, away from her and our daughters. Once I graduated with my master's, I thought I could get an industry job and we could move on as a family: delayed gratification. She needed to hold on. Once again, thoughts of my choice "To Be or Not To Be" dominated my mind.

Audine's parents were teaching school in Colorado when they called to tell us her sister Carol was seriously ill—she had a brain tumor. They were preparing to drive her to a hospital in St. Louis where the doctors could hopefully help her. They would stop by Manhattan on the way.

We waited anxiously for the mid-afternoon arrival. Carol was as sweet as ever, but she didn't look right. She hugged each of the little girls but had difficulty speaking and walking. They stayed with us for an hour before heading back toward I-70 and onward to Missouri.

Carol died about a week later, on May 29, at age sixteen. She was sitting up in the hospital bed; she and her mother were talking when Carol closed her eyes, stopped talking, and quietly passed away.

The funeral would be in Kansas, near her father's ancestral home in the Flint Hills—the same day as the Advanced Quantum Mechanics final exam. Should I go with my family to Carol's funeral, or stay and take the exam? In the end, I stayed. My exam results were good, but I would regret this decision for the rest of my life.

Rocky, Glen and his girlfriend Lee Ann had become a threesome. I enjoyed participating in their back-and-forth banter. They made trips to a roadhouse east of Manhattan where Glen sang and played his guitar. I wanted to go along with them, but I chose to divide my time and energy between research and my family. One morning Rocky came in looking a little bleary eyed. "What happened?" I asked.

"We came close to totaling the car; took a curve too fast and hit a telephone pole. Glen's knees hit his guitar, smashing it to pieces."

"Is everyone all right?"

"Yeah, we're all okay," said Rocky. "I had three girls in the back, and they were okay as well."

Around that time, Lee Ann's allegiance seemed to have shifted toward Rocky.

Tom was another student; a year behind me, blond hair, and full of energy. We both spent a lot of time in the physics machine shop. Before long, we were friends. We had him over to our apartment a few times. Tom knew his way around tools in the shop, but he was also prone to risky behavior. On one memorable occasion, he decided to make some nitrobenzene inside a lab fume hood. It blew up, nearly destroying the hood.

Meanwhile, I was searching for a way to display my electron density maps in three dimensions. My program had taken a bit less than seven hours on the IBM 1410 to compute forty layers through the three-dimensional function, but I had no way to visualize it. Accordingly, I built a wood frame holding forty glass sheets standing on edge, spaced a half-inch apart. I laid each twenty-four-by-thirty-inch sheet on top of my printed contour map representing a layer of the three-dimensional function, tracing it using a wax crayon. The panes were illuminated by fluorescent tube lights on their edges. Toggle switches selected one or more layers through the three-dimensional structure. The electron density maps glowed brightly because the wax interfered with the total internal reflection of the light in the glass. All of this took a lot of time working in the shop, which meant time away from Audine and my daughters. But at last, I could gaze into the full 3D structure of crystalized sodium hyponitrite computed by my program.

It was late July, and I was writing out my thesis document. But I decided I wanted another way to display my crystal structure. The location of atoms in the structure could be inferred from the peaks on the glass sheets of the 3D electron density, but I wanted to directly represent the sodium, nitrogen, and oxygen atoms with a ball-and-stick model.

Glen and Rocky worked with me to make this model of the sodium hyponitrite crystal. Atoms were represented by painted cork balls. We mounted the balls with holes drilled through them on vertical glass rods held between horizontal sheets of Plexiglas. The rods were equipped with spring-loaded rubber feet, enabling the cork balls to be stationed anywhere within the volume defined by the model. Bonds between the "atoms" were provided by telescoping spring-loaded brass tubes. I spun-formed the ends of the tubes on a lathe using a variation of the method we had used to make the tail cones for our Arcturus rocket.

While I worked on my model, Tom was working on something in the shop. He was busy welding some pieces of angle iron to make a rectangle. "What are you making?" I asked.

"I'm making a small table for Audine," he said. This struck me as odd.

"Oh, really? She had told me she needed something. I was going to make her a table out of wood," I said.

"It's no big deal. This won't take me long," he said.

I was so proud when I graduated with an M.S. in physics in August 1965. For the first time, I felt I had become an integral part of the K-State Physics Department, not merely a student. Rocky and Glen got their M.S. as well. As experimentalists, we would be required to pass a tough exam to go on to the Ph.D. program.

While preparing for the big exam, to be held in the first week of September, I was so enamored of our crystal structure modeling system that I spent some extra effort—just for fun—modeling the structure of vitamin B12, the largest and most structurally complex vitamin, whose crystal structure was already known. It's comprised of carbon, cobalt, nitrogen, oxygen, and phosphorus atoms. This took some effort to do, but the result was simply amazing. I could walk around the structure to appreciate its beauty from different viewpoints.

118

It was a tough exam designed to weed out nearly half of the M.S. students aspiring to go on to the doctoral level. I managed to pass, but Rocky did not. "It's been great working with you, Ralph," said Rocky as he handed me a pencil sharpener and paper punch.

"What's this?" I asked.

"Going away present. I got a job at a company called Autonetics, in southern California."

I needed only a few more classes for my Ph.D. requirements, then I would find a new research project to work on. Only about three more years of effort would be needed to complete my original contribution to physics. I knew that my goal was within reach.

It was an unusually warm October; the trees on campus turned into a lovely profusion of orange and red. Our marriage had also taken a turn. Audine was sitting on the edge of the bed when I entered the bedroom. She looked up at me and said, "Ralph, I'm divorcing you."

"What?" I said, stunned to the core of my being. "Why are you saying this?"

"I just can't take it anymore. I no longer love you."

Anger and dismay welled up in me. "What about the girls?" They were asleep in the next room.

"They will stay with me," she said.

My mind rushed ahead of me—angry and upset; arguing back and forth; what about our marriage vows; my total faithfulness to you, the first and only woman I had ever loved; no one in my family had ever been divorced; working hard to support my family, put food on the table; never drank so much as a single beer; a responsible husband and father; what's so wrong with me; why are you doing this? *What will I do?*

No amount of begging and pleading made a difference. A week later I came home to an empty apartment. Audine and the girls were gone. There was no note, only my own possessions remained. Weeks went by as I struggled with my class in advanced quantum mechanics without a clue as to where they had gone. I called her parents, but they didn't know.

I burdened Don with my anguish. "I think I'm going crazy with grief."

Reluctantly, he started to clue me in about Tom. "They're in a house a few blocks east of campus." He gave me the address. Then he said, "Your daughter Dana fell out of the second story window and spent a few days in the hospital."

It was an old house in a neighborhood of old houses. Glancing upward, I saw the window my two-year-old must have fallen out of. I thought my heart would beat through my chest as I pounded on the door. Tom opened it, Audine behind him in shadow. Blind with rage, I lunged at him. He turned swiftly and ran out the door. I ran after him as fast as I could, cursing at him and chasing him down the sidewalk for half a block. I couldn't catch him. If I had had a knife or a gun I am sure I would have spent the rest of my life in prison.

Our girls were okay. Apparently Dana or Carol had pushed a screen open when she fell out. I berated Audine for being so careless and for not even notifying me that our daughter was in the hospital. More arguing and yelling ensued. I couldn't believe the woman I thought I loved would do this to me.

My mind was swimming as I nursed my sore legs for a week. I failed the mid-term exam in advanced quantum mechanics; couldn't go on like that so I dropped the course. Applying for jobs, I got an interview with Texas Instruments in Dallas. With a secure paying job, I hoped Audine and I could get back together.

I boarded the DC-3 at the Manhattan airport, flew to Kansas City, and on to Dallas in a four-engine Lockheed Electra turboprop. I did my best through multiple interviews, feeling like a beaten man. A few days after returning I got the news: there was no job for me at TI.

* * *

On the first of December, Dana had turned two and Carol was four. My mind was made up when I returned to the old house. I was crying when I said goodbye to the girls and Audine. "Promise me you will wait a year before divorcing me. I'll try to get a job in California where I can use my master's degree. Things will get better for us." Audine was silent. My girls were too young to understand what was happening.

The dream of getting a Ph.D. in physics was over. I dropped out of graduate school, packed up my clothes and drove to my parents' home in Wichita.

III
ANAHEIM

15 California

I T WAS EARLY MORNING in Wichita when I loaded up my car and headed west on highway 54 with my Rand McNally and $200 in cash. On advice from Dad, I had traded my old Ford for a 1956 two-tone green-and-white Chevy. Three weeks after leaving Manhattan, fall was turning to winter, and the weather had turned clear and cold. My head was filled with storm clouds.

After Rocky had failed to pass the Ph.D. qualifier exam, he had moved to Anaheim, California, accepting the job at the Autonetics division of North American Aviation. I thought I might also get a job in California, perhaps even at Autonetics. Rocky would take me in as a roommate. It was an impossible dream, but once I was earning some money, Audine and the girls could join me in California.

* * *

Driving through Kingman and Pratt, I stopped in Greensburg where, years ago, Dad and I had visited the "World's Largest Hand-Dug Well," and the thousand-pound meteorite. Turning northwest, passing through Dodge and Garden City, the landscape turned almost mathematically flat, and I thought of our many trips to western Kansas in the Tri-Pacer.

I crossed into Colorado at dusk, passing through small towns, each with grain elevators rising like castles above the dry plain. It was dark when I pulled into Walsh, where Audine's parents had taken teaching jobs in the small school. They were happy to greet me and were full of questions about their daughter and granddaughters.

After dinner that evening, I pressed the case for reconciling with Audine. They were both sympathetic, expressing a desire for my family to be reunited. I hoped their influence would help.

We got up before dawn. My father-in-law drove me a short distance to a lake where thousands of migrating birds had gathered. He talked about what an incredible sight this was in the pre-dawn light. But I was in no mood to enjoy such a spectacle. Then he said, "Well, you know you can lead a horse to water, but you can't make it drink. Eventually, if you leave the horse alone, it will take a drink. I am sure she will come around."

I thought about it for a while as I gazed out over the lake. Finally, I replied, "I don't think so."

* * *

Driving south over Raton Pass into New Mexico, I found a cheap motel on the outskirts of Santa Fe. The next day's driving west put me through Albuquerque, where I picked up highway 66. In 1955, I had traveled with my parents and siblings on the El Capitan that had run on the rails nearby.

Crossing into Arizona, near Lupton, I was once again running out of gas and getting low on cash. I stopped beside a massive sandstone cliff at the Chief Yellowhorse trading post, full of Indian curios and souvenirs. The young attendant, dressed in faded greasy overalls, topped up the Chevy, and then asked me, "Where you headed?"

"California," I said, with a hint of enthusiasm. The rubber band that had been trying to pull me back to Kansas was stretching thin, and another rubber band had taken over, pulling me west toward Los Angeles and my future. "I am trying to get a job there."

"Oh, yeah. What do you do?"

"I'm a scientist," I said, feeling somewhat confident.

He replied with a loud, "Ha!" and started laughing his head off as he put the nozzle back in the pump.

* * *

At last, after Flagstaff, I crossed the bridge over the Colorado into California. Abruptly, the landscape turned from desert to a lush green. But the route soon turned back to desert as I drove past Amboy Crater and into Barstow.

I had telephoned Rocky before departing Wichita; he had given me detailed driving instructions which I had carefully written down and put safely in the glove compartment. Rocky lived on the first floor of an apartment building wrapped around a swimming pool.

It was evening when I pulled into the tree-lined street and parked. I was tired, hungry, and nearly out of gas. I had arrived in Anaheim armed with a master's degree in physics, a ten-dollar bill, and a pocket full of coins.

Rocky's door was open. "Hey Rocky, I made it!"

"Hi Ralph! It's great to see you again." I stepped into the apartment, and we gave each other a big long-lasting grad student buddy hug. "Here you go. How about a beer?"

I had so far shunned any form of alcohol, since I knew it destroyed brain cells and I didn't have many to spare. "Okay, I'll give it a try." We spent the evening talking about our physics student friends, Audine and the girls, and life.

16 Autonetics

T HE NEXT MORNING, I drove over to the Autonetics office on Kraemer Boulevard. I was offered a job and I took it — at a salary of eight thousand dollars a year — much more than I had made as a graduate student. I was hired into the Systems and Reliability Engineering Department of the Navigation Systems Division, part of North American Aviation. I can't believe it was so easy to get a job.

It was thrilling to be a part of the famous company that had built the historic P-51 Mustang fighter, the B-25 Mitchell bomber, and the F-86 Sabrejet. I knew work was underway on the XB-70 supersonic bomber, and the Apollo command and service module which would someday take men to the moon.

* * *

Autonetics was an array of white buildings near the Riverside Freeway created by pouring reinforced concrete in large slabs, and then tilting them up to form walls. A lot of secret work was going on inside, so there were few windows. I was assigned to the upstairs floor of Building 72, which was surrounded by a high fence.

With my new white shirt and black tie, I showed my badge to the guard. Nearly the entire floor, of perhaps 10,000 square feet, was divided into a bullpen of modules with five-foot high partitions. The exceptions were offices for managers along each outside wall. I learned that the name "Autonetics" was a contraction of "automatic" and "cybernetics." I thought that was pretty cool.

Systems and Reliability Engineering had three main groups, focused on navigation issues in airplanes, submarines, and missiles. I spent time talking with people in each group, and in the end was given my choice of which to join. A nice fellow ran the airplane group. Barbara Conrad ran the submarine group. I was told she was onboard the *U.S.S. Nautilus* during her historic voyage under the North Pole. Dick Snyder ran the missile group. He was a man on a mission. I picked the missile group because I needed discipline and thought he might be the hardest to work for. And if things went well, someday, I might work on rocketry.

I was assigned to a ten-person module, two rows of desks with five persons in each row. My desk was on the left in the back, so I could see everyone in front of me but the others wouldn't notice me without turning around. Someone gave me some paper and pencils. The pencils had the word "PRIDE" on them, alongside the Autonetics logo: a squiggly "A" resembling an oscilloscope trace. I asked why "PRIDE," and was told it stood for "Personal Responsibility In Daily Effort." I was not allowed to do any work until my secret clearance came through. Mostly I sat and doodled for several weeks, but nevertheless got paid.

My first job was working for a Ph.D. physicist concerning the effect of magnetism on a new type of gyroscope that had no moving parts. The so-called ring laser gyro had been invented at Sperry Gyroscope Company only two years before. In this device, beams of light travel in clockwise and counterclockwise directions between three mirrors arranged in a triangle. The beams interfere with each other at a detector producing a signal related to the rate of rotation in the plane of the entire device, something called the Sagnac effect. The absence of a rotating flywheel might make it ideal for use in a missile, if its accuracy could be proven to exceed that of a well-made mechanical gyroscope.

The invention suffered from several limitations, one being its sensitivity to magnetism. Even a very slight error in the gyro, perhaps caused by variations in the Earth's magnetic field, would be disastrous to the flight of a missile. The physicist in charge had written down several pages filled with equations describing his theory of the effect. My job was to solve the equations by creating a program in the FORTRAN language, using an IBM 7094 computer.

The physicist in charge continually pressed me to work faster, sending me new equations to program each week. He was worried about being fired, telling me he would have had a happier and more fulfilling job as an insurance salesman. He didn't have the time to explain the actual physics to me in terms I might be able to understand, so I never actually knew what I was doing, but I was glad to be distracted from worrying about Audine and my girls.

I never saw an actual ring laser gyro because of a pervasive rule called "Need To Know." All I needed were the equations. There was no need to see the actual secret device. This was disappointing. On the other hand, I was learning more about FORTRAN and the 7094—by far the most powerful computer I had ever worked with.

The 7094, located in the Data Center a quarter mile away, used large numbers of transistors in lieu of vacuum tubes, and had 32,768 words of magnetic core storage instead of a magnetic drum. Each of the words had 36 data bits. The core memory stack had 36 layers of bits, immersed in a temperature-controlled oil bath. The 36-bit words were conveniently considered as twelve 3-bit segments. Each segment thus could represent numbers from zero to seven, so words could be expressed in a base 8, or octal, format. With a two-microsecond basic cycle, the computer was blazingly fast.

My usual routine involved translating the equations into the FORTRAN language, combining them with numerous other statements dealing with branching, looping, and input/output. I used my pencil to write the statements, line by line, onto FORTRAN coding forms, each line representing a single 80-column IBM punched card.

When I thought I had written enough of the program, I took the filled-out forms to the keypunch operators. The operators, mostly young women, read my coding forms and punched the cards for me on their IBM 026 keypunch machines. The cards' upper left corners were cut off, so they could be stacked correctly to form a deck.

The punching process took a few hours, so I would trade fluorescent lights for a few minutes of California sunshine. I'd walk around or get something to eat from the "garbage truck" parked outside. The food, mostly simple chicken salad sandwiches and the like, was cheap and tolerable. The man running the truck made it a point to learn everyone's name. Soon, he was able to greet me by name with a witty remark.

After I got the cards back from the keypunchers, I scanned them for errors by reading the printed characters along the top edge of each card. Usually, I found a few mistakes, so I brought them back and joked around with the operators while they were being re-punched. I particularly liked talking with a cute blonde girl named Georgie. When I had finally built up my program deck with several hundred cards, I added some special cards at the beginning that were not FORTRAN but were instructions to the so-called IBSYS program which ran the computer. These cards told the 7094 what to do with the remainder of the deck.

Every few hours, someone would come by picking up decks from my colleagues and me to be carried off to the 7094. In another few hours, my deck would come back, along with the results. The results were printed on fifteen-inch-wide fan-fold paper with light blue bars, making it easier to read. After fixing errors, the whole cycle would repeat, day after day, until the results were satisfactory.

129

I stacked up old printouts and program listings on the floor next to my desk. In a week's time, the stack would grow to a height of a foot or two. Someone would come by at night to pick up everyone's stacks and old card decks, dumping them into big bins out on the loading dock. Once a week, late at night, a large truck with an enormous canvas bag on top would pull up to the dock. Inside was a huge roaring paper shredder which would spend hours chewing through all the paper. Before dawn, it would drive away. All my hard work ultimately ended up as tiny scraps of paper.

* * *

Four months into my job, I was starting to get the hang of it. At the apartment, Rocky and I were getting along fine. He was a great help during our many talks about the loss of Audine and my girls, stoically putting up with my moaning and groaning. His job, in a different building at Autonetics, involved developing secret surveillance capabilities, which he could not discuss in any detail. Swimming in the pool after work was a great joy, as were trips to nearby Laguna Beach.

A bar adjacent to Autonetics featured topless Go-Go dancers. Rocky and I went there a few times. One Vietnamese girl looked nice to me. Sometimes I felt lonely as I nursed a single beer for the whole evening while watching her.

At another topless bar, in LA, we were joined by Rocky's friend Jerry, an unrefined kind of guy—quite a character. A man with a flashlight inside the door asked me for my driver's license. When I took out my wallet, the cards fell on the floor. Groping around in the darkness, I began picking them up mostly by feel. When I finally had them and stood up, my face brushed against the exceptionally large naked breasts of a tall dancer standing directly in front of me. I felt a flush as heat rose in my cheeks, but I showed her my license and she let me in.

My old friend Phil Roberts, from Propulsion Research, drove down to Anaheim with his new wife Joan for a joyous reunion. Phil was now a Ph.D. student in physics at Caltech in Pasadena. After their visit, I made quite a few trips back and forth to catch up on our shared history.

* * *

By now, I was getting acquainted with the panoply of rather strange people surrounding me. Most were under thirty and had excellent programming skills. I was rather shy and reticent to tell others much about myself, especially that my family had left me. I was too hurt and ashamed. Better to let them wonder.

Lucille Randolph was three years older than I. Intelligent, she was a master programming whiz. Her lilting accent hailed from the Louisiana bayou country.

Ever willing to help, she was quick to give me tips and pointers on how to get things done. I learned later that she had earned a double major in math and physics, but the physics jobs she applied for all had specified "men only." A counselor told her to ignore the physics part and apply only for math jobs. That's how she landed the job at Autonetics.

Sonya Veilleux, a tall striking redhead was quick to tell me she was tired of my habit of staring at her. I lied and told her I was staring into space thinking about the solution to a problem.

Once in a while I interacted with other women in different cubicles. One told me her dream job was to be a madam running her own whorehouse. I wished her the best of luck. Another told me about eating a lot of beans, farting in the bathtub, collecting her farts in a glass underwater and then lighting them on fire. "You should try it," she told me.

Gary Roberts was one of the people in my cubicle. He impressed me with his studious focus on work and intelligent conversation. He was working to create 3D math models for radiation shielding. Later on, he worked on a real-time computer language he called Reticol, writing the code in IBM's new PL/I language.

Bob Moffitt sat a few desks in front of me, a high-school dropout who had spent four years in the Marines before earning his GED, then returning to get his high school diploma by taking night school classes. Bob had joined Autonetics, working in the mail room. Now he was pursuing an associate degree part time at Fullerton Junior College.

Al Sheue, a thin wiry kind of guy, had strong opinions about most everything. He kept proselytizing that everyone was programming the wrong way; instead, he felt we should use his "inside out" programming mantra. Conventionally, one writes a whole bunch of preliminary code to read data into a program, then executes the important nugget of code, then follows it with a bunch of code to output the results. This paradigm hides the important part in the middle of all the cruft. Al did it the other way around. At the highest level, you only saw what was important. He would push the input and output down to hidden lower levels.

I was awed by wizard programmer Jerry Brown, author of the Magic and Merlin assemblers. An assembler is a program for converting a source program written in a way that is understandable into the arcane machine code of the computer—like the assembler I had used years ago on the Bendix G-15. Jerry would sometimes make minor changes to a program by taping chaff into selected holes in the binary cards to avoid the hassle of sending his program to the Data Center for re-compiling. He dug the chaff out of the waste bins near the card punch machines. Someone told me he hadn't cashed his paychecks for years.

* * *

It was late in the afternoon of June 8, 1966, when I was beginning work on a new project. My thoughts were interrupted by an announcement broadcast over the PA system.

"May I have your attention, please? May I have your attention for this announcement." Then a pause as we all looked up from our desks. "The XB-70 has crashed north of Barstow, with loss of life. I repeat, the XB-70 has crashed. No further details are available at this time."

All around me, stunned silence. Then everyone got up, put away their work, and started heading for the doors.

A few days later, we learned that our giant Mach-3 supersonic bomber had been flying in a formation with four other jets in a photography shoot for General Electric, maker of the engines of all five aircraft. One of the four planes, an F-104 Starfighter, had collided with the XB-70's right wing. Joe Walker, flying the F-104, was killed instantly. Al White, pilot of the XB-70, was able to eject, surviving with serious injuries. Carl Cross, co-pilot of the XB-70, was killed after the bomber entered an uncontrollable spin and crashed.

* * *

My sadness about the XB-70 crash melded with my grief over the breakup of my marriage. I couldn't get these thoughts from my mind as I lay awake in bed. The situation was strange. Rocky was happy and looking forward to starting a married life with Lee Ann; I was sad because my marriage was ending.

Later in June, I arranged for Audine to fly out to California so we could talk things over. Since I didn't have a place of my own, I put her up in a motel on State College Boulevard. Rocky had gone back to Kansas for the wedding, so Audine and I met briefly in the apartment. As we talked, a small earthquake caused the water in the pool to shimmer. *Maybe the gods are angry.*

The next day, I argued my case for reconciliation. I declared my love for her and told her how much I missed Carol and Dana. I wanted to hug her but thought it would only serve to push her away. Leery of displaying too much affection, I thought she might think I was faking it. In the end, she told me once again that she no longer loved me, and that life was too short to keep on with the marriage.

I drove her back to LAX, and it was over. On the way back to the apartment, traveling on the Santa Ana freeway, I felt my loss as a physical pain in my chest. People talk about "heartache." I guess it's what I had.

Since Rocky would soon return with Lee Ann, I had to find a new place to live. I moved into a cheaper upstairs apartment without a pool, a half-block nearer to State College. I made a wedding present for the newlyweds by drawing an ink sketch of John Kennedy on a piece of white cardboard, based on a photo from *Profiles in Courage*.

* * *

My new job was analyzing gravity measurements taken from a ship at sea. The idea was that a nuclear submarine could precisely measure the pull of gravity as it cruised below the surface, using this information to correct its inertial navigation system.

A ship out at sea had an Autonetics SINS (Ships Inertial Navigation System) on board, the same type that navigated the *USS Nautilus* on its famous voyage from the Pacific to the Atlantic under the ice surrounding the North Pole. Our ship would sail back and forth in a regular pattern on the ocean's surface taking readings from the SINS, which were recorded on magnetic tape.

A large truck, with another SINS mounted inside it, periodically drove around Southern California to test its navigation accuracy. A laser beam from the truck bounced off reflectors fastened to telephone poles to mark when it passed by. Survey crews had precisely located these known landmarks.

I actually saw one of these huge SINS, in a lab on the first floor of my building. The machine towered over me, hanging from a strong metal frame. The heart of the system was a "stable platform" attached to ring-shaped gimbals allowing the platform to tilt forward and backward as well as left and right. The gimbals were basically like the hinges on a gate, only made with much greater precision. Gyroscopes, mounted on the platform, spinning at a terrific speed, kept the platform level despite rolling and pitching motions of a ship. Three sensitive accelerometers, also mounted on the stable platform, measured accelerations in three different directions.

By continually adding up acceleration information, the velocity of the ship could be computed. Likewise, by continually adding up the velocity information, the ship's global position could be determined. It seemed like magic, and maybe it really was a sort of magic.

Every week or so, I would receive the tape from the ship containing all the data from its SINS. My task was to make sense of this information, particularly focusing on the vertical component of acceleration along the direction of the Earth's gravitational pull. The actual gravity values, however, had to be teased out of all the ship's motions because of waves, ocean currents, and tides. We were looking for changes in gravity as small as one part in a million, which might be caused by undersea ridges or canyons that could serve as navigational waypoints.

The first problem I had to face was reading the tape into the 7094. The tape recorder on the ship magnetized a tiny stationary strip of seven bits across the tape's half-inch width, then a fraction of a second later the tape was stepped forward a small distance, about two-hundredths of an inch, and the operation would repeat. So the tape was not in motion during the time data was being written. The recording would go on 24 hours a day for a week or two without interruption, on a tape that could be nearly a half mile in length. Thus, when the tape got to me, it would contain thousands of unbroken feet of recorded data from the SINS.

The 7094 had ten IBM 729 magnetic tape units. I naively thought I could have the operator mount my SINS tape on one of these units, and the computer could simply read it into its memory for processing. This was not the case. A 729 could only read tapes which every so often had breaks in the data, critical so-called "inter-record gaps." My SINS tape was devoid of inter-record gaps. The 729 unit read tape at 75 inches per second but needed a three-quarter inch gap every so often to allow the mechanism time to stop. The data could then be processed, and afterward the next record could be read, and so on. Like drinking water: you open your mouth, fill it with water, then close your mouth and gulp it down. Only then can you take another drink.

I was stuck on this problem, seeking the advice of veteran programmer Al Sheue. He told me the only hope for reading the tape was to use a technique called "double buffering." I needed to read the tape into a first memory area, called a "buffer," then when the buffer is full—without stopping—swiftly switch to reading the tape into a second buffer. While the second buffer is filling up, data in the first buffer could be written out to another tape, followed by an inter-record gap. The roles of the two buffers could then be quickly swapped, and the whole process repeated until the new tape would contain all the SINS data, but with the essential gaps. It would be like having two mouths in your head. You could drink with one mouth while swallowing with the other.

I could not do double buffering from the FORTRAN language, so had to program it in assembler language, one step above the 7094's machine language. This was similar to programming using Intercom 500 on the Bendix G-15, much like I had done for my rocket calculations back at Beech Aircraft.

Writing the SINS data on the output tape was easy for me, since I had done this sort of operation before on the IBM 1410 at Kansas State as part of my master's work. Once all the data was safely written, the tape could be rewound and read into my program for processing—hopefully yielding the gravity information.

I worked with a mathematician who had devised a "triple integral filter" with the potential to eliminate nearly all the extraneous ship's motion, leaving behind a faint remaining signal, the vertical component of gravity. After working on this for the good part of a year, we began to observe results corresponding with known features on the ocean floor. This was exciting! As a bonus, I was able to plot graphs of the results using the Stromberg-Carlson 4020 printer, which was installed near the 7094.

The SC4020 had a cathode ray tube with an internal mask stencil through which the electron beam is deflected to choose a symbol, and the symbol-shaped beam is then deflected to a target position on the faceplate of the tube. A film camera captures the imagery produced on the faceplate. The film is then developed chemically in a lab and later enlarged as a paper print. Output from the SC4020 was available on a twice-daily basis. My mathematician colleague was happy to get the results in graphical form. It was my first foray into what would eventually be called *computer graphics*.

As interesting as this work was, we knew it was likely that many decades would pass before it would actually become feasible for submarines to navigate, without surfacing, using gravity measurements to periodically correct their inertial navigation systems.

* * *

"Here it is," Rocky said, as he handed me the thick envelope. He had called me over to his and Lee Ann's apartment when the letter came in the mail. My divorce papers were inside. Amidst the legalese of this final decree, were the words justifying Audine's reason for leaving me: "EXTREME MENTAL CRUELTY," spelled out in capital letters. This pronouncement shocked me. I didn't have the stomach to travel back to Manhattan to contest it in court. It stipulated a requirement to pay monthly child support, but thankfully no alimony.

I had been scrimping to save money. I had still held out a sliver of hope that Audine would come to her senses, but now I decided to spend a little of my savings on myself. I had seen an expensive wooden chess set in the window of a downtown store, so I impulsively returned there and bought it, hoping it might salve my wounded soul.

* * *

A married colleague at work had taken pity on my situation and put me in touch with her single friend. I drove up the Santa Ana freeway to her apartment in

North Hollywood. Frankie opened her door—dark hair, beautiful blue eyes, and pale white face. She was English, or rather, grew up in England before her family moved to California. She invited me in and we shook hands.

"Hello, nice to finally meet you," she said. We spent the evening sitting on her couch, recounting our histories, and relating our childhoods. She told me about growing up in London. "When I was small, we took refuge in the tubes to get away from the bombings. I remember that time well."

"I remember stomping on tin cans collected for the war effort," I said. "Meat stamps, no metal or rubber. Toys were made from glued-together wood and cardboard. We made toys from the cardboard separators that came in boxes of shredded wheat. Mom made me a toy motorcycle from a pair of discarded furniture castors."

"I remember the shredded wheat boxes—and the cans," she said, with a delightful laugh.

"You know, when I was a student at Kansas State, I met a German classmate from Berlin who told me he also stomped on cans. He also told me that after the war the kids would go around collecting the skulls of dead people, getting a reward for each one they turned in," I said.

"How odd. Three little kids in three different places in the world, all doing their best to help their war efforts," she said, with what I thought might be a brief flicker of sadness. I began to bond with Frankie, asking myself if she is a woman I could fall in love with.

From this evening on, I would make the long drive to her apartment on occasional weekends. We would take in the sights—Mulholland Drive, Griffith Observatory, La Brea Tar Pits. A few times we would take drives toward Santa Barbara, once to the beach at Carpinteria. Several months into our relationship, I met her parents who lived in Santa Ana. Curiously, her mom took me aside and confided to me her daughter—who usually wore long pants—had varicose veins in her legs. I was not sure of the significance of this information. Maybe her mother didn't want me to be involved with her and was trying to scare me off.

17 Incident at George

M Y FRIEND Steve Hawkins, one of my main rocket pals and fellow poet from high school, had recently returned to George Air Force Base from his deployment at Ubon airbase in Thailand. George AFB, out in the high desert near Victorville, was seventy-five miles northeast of Anaheim. I decided to visit him there, to see him for the first time since he'd driven from Wichita to Manhattan to visit me on November 22, 1963, the day President Kennedy was assassinated.

As an Airborne Electronics Navigation Equipment Technician, his job was to go out to the flight line to work on specific planes, such as F4C Phantoms, operating over North Vietnam. Each plane had a logbook kept in the cockpit with entries telling about problems (squawks) and the pilot's complaints. It was Steve's job to verify each complaint, if possible, and rectify them. While in Thailand, Steve was promoted to Airman First Class.

I started out with a California road map on a clear sunny Saturday. Heading north on State College Boulevard past the strawberry fields, I swung on to the Riverside Freeway, and shortly transitioned to the northbound Orange. In an hour, I headed north on I-15 over the Santa Ana mountains, and after another hour into Victorville. The famous cowboy couple Roy Rogers and Dale Evans were living in the tiny adjacent community of Apple Valley. As a youngster growing up in Wichita, I enjoyed watching their western movies at the Tower Theater. Turning into the entrance road to George, the landscape was sand, punctuated by a few desert scrub. Steve had given me his barracks number, so I told this to the guard, and after a minute or two he waved me through.

We met in his room, which he shared with another airman. The room was drab and bare.

"Hi Ralph, how many years has it been?" said Steve. It was a joyous reunion. Steve was the same tall, lanky, scholarly person I had known before — perhaps a bit thinner. Over a beer, we filled each other in on personal things that had transpired since our last meeting. Steve told me of his time at Ubon and some wild stories about the Air Force guys in Thailand. I filled him in on my divorce from Audine. He asked about the work I was doing at Autonetics, but I couldn't tell him about it.

"How is your faithful car?" I asked.

"I loaned the fifty-five Plymouth to a buddy who smashed it into a rock, totaling it."

"Oh, gee. That's too bad."

"I replaced it with a fifty-four Lincoln, my huge pink tank of a road machine," he said with a grin.

Toward evening, Steve wanted to show me some Phantoms out on the flight line. He drove us in the Lincoln a short distance to the tarmac, parking a hundred feet away from a row of F4Cs with brown and green camouflage markings. It was such an awesomely deadly warplane with its distinctive negative-dihedral tail and weapons pylons.

We stepped out of the car and Steve cautioned that we could not cross the white line painted on the concrete in front of us. I started asking about some of the features of the Phantom. "What's that thing sticking out under the nose?"

Before Steve could answer, two vehicles rushed toward us on our right, skidding to a halt twenty yards away. A half dozen uniformed and helmeted air police jumped out, leveling their M16s directly at our heads. "Halt! Put your hands in the air!"

Trembling, with arms raised, I turned to Steve, "Are those machine guns?"

"Uh," and after a pause, "Yes."

It turned out to be a long night. We were herded into a small shack for questioning. Steve was asked his rank and serial number, which he readily answered. We both were treated to a litany of probing questions.

"What's the problem, sir? We didn't step over the white line," said Steve.

Gesturing toward the Lincoln, "Your car back there is two feet over the yellow line." He turned to me, "What's your rank and serial number?"

"I'm not in the Air Force, but I'm actually working for the Air Force at Autonetics in Anaheim." I was careful to emphasize, after a pause, "So, we are on the same side." Then, I thought maybe I sounded too smart-alecky.

While Steve was in Thailand, the boundary at the George flight line had apparently been moved back. We had crossed it. Since bombing had recently commenced in North Vietnam, protests against the war had gained steam.

Increased attention was paid to base security, and the air police were ready and willing to use deadly force to protect the warplanes from any kind of sabotage.

Phone calls were made, and we waited in silence to find out what would happen. Many hours later, with suitable admonishments, we were set free into the night. In the end, neither Steve nor I suffered any consequences. Years later, Steve told me the only reason he was not thrown in the brig was that my security clearance was even higher than his!

Driving back home that night through the desert and lost in thought, I knew it would feel good to get back to the safety of Anaheim and Autonetics, where soon I would be involved in an exciting new project.

18 Minuteman III

S Y ALPER WAS IN CHARGE of a row of giant electronic racks bulging with hundreds of wire cables connecting everything together. Standing in front of it, I was reminded of the patch plug panel for the IBM 610 computer that I had worked with years ago, only this thing was massively larger. Alper was a large, pleasant man whom I had met several times before. He usually wore a dark blue suit and black tie. He had some kind of issue with his heart and talked frequently about his efforts to lose weight.

"This is the analog computer we use to simulate the flight of a Minuteman missile," said Mr. Alper. "It solves the differential equations of motion which govern the entire system." I had learned something about analog computers from my efforts to solve the Schrödinger equation of quantum mechanics back at Kansas State.

"It will need to be considerably upgraded to simulate the flight of the new Minuteman III," he told me. I had heard about the first Minuteman silo launch tests at Edwards AFB from Phil Roberts in 1959, the year I graduated from high school. Minuteman departed radically from the Titan II ICBM. It was much smaller, cheaper, and most importantly was based on solid propellant rocket motors, like the rockets I had built. It could rest quietly in underground silos for months or even years with little maintenance. The first Minuteman became operational in 1962.

Minuteman II came along as an improved version; its production and deployment began in 1965. With more powerful rocket motors, it had increased range, and greater throw weight. The guidance system was also improved, yielding better accuracy over a wider range of possible targets. Minuteman II carried a nuclear warhead with a yield of a bit more than a million tons of TNT, much weaker than the nine-megaton warhead of the Titan II—but still about eighty times greater than the destructive power of the bomb that had leveled Hiroshima.

The Soviets were actively developing anti-ballistic missile technology for defeating Minuteman. To counter this, some of the Minuteman II missiles were being equipped with decoy warheads and other so-called penetration aids to confuse enemy defenses. But the ultimate goal was a new version of Minuteman that would be an even better weapon. Sy told me, "Minuteman III will have an additional fourth stage based on liquid propellants. It will also carry as many as three warheads."

Dick Snyder called a meeting in his office. Sy Alper, Lucille Randolph, Dick Capello, Gary Roberts, Al Sheue, and several other senior people were there. Mr. Snyder laid out some of our plans for Minuteman III. "It's the first Minuteman to have MIRV capability. MIRV stands for Multiple Independently Targetable Re-entry Vehicles." He continued to describe the missile currently under development, and the role Autonetics was to play in its creation. "One thing we have decided, is we should make an all-digital simulation of Minuteman III. We need to simulate the flight from liftoff to impact of the RVs. Is this something anyone would like to take on?"

I'm not sure how I got the courage to pipe up, "I created some digital simulations of my amateur rockets. One was a two-stage rocket more than twelve feet tall." A brief silence greeted me as everyone glanced in my direction.

"Go on. What computers did you use?" asked Mr. Snyder. I haltingly described my work in some detail as everyone listened, but soon started to regret speaking at all. Recalling my first primitive steps with the IBM 610 computer, and later with the Bendix G-15, and all of the work simulating the flight of the Cryolite rocket, I felt foolish talking about it in front of my senior boss and the others.

* * *

A couple of days later, Mr. Snyder called me into his office. Sy was there as well. "I think you should lead the simulation effort." The realization hit me that he was forcing me to make an incredibly important decision. A wave of adrenaline coursed through my body as fear welled up in my throat. I could feel a few beads of sweat forming on my forehead. I had to make a decision right then that might have a profound impact on my future. If I declined, the opportunity to do something great at Autonetics could be gone forever. If I accepted, there was a good chance that I would fail. My mind was too slow to weigh the benefits and costs of my choice, but I knew I would have to cope with the consequences. I thought about it for a few minutes, and at last agreed to do it.

Sy handed over a huge stack of papers describing the new Minuteman III in detail, along with hundreds of pages of equations governing his analog simulation of Minuteman I and II.

Back at my desk, fear gave way to excitement as I began to pore through the mountain of material. Boeing Airplane Company had overall responsibility for Minuteman. I learned that Autonetics was responsible for the guidance and control (G and C) for all three versions of the missile, but I did not fully understand what that meant.

Drawings showed the beauty of the missile's design. Finless and sleek, it reminded me of a gigantic 30-06 rifle shell—wide at the bottom, then tapering to a narrower profile, and topped with a perfect ogive nose cone, like the ones I had carved out of balsa wood.

Thiokol made the missile's powerful sixty-six-inch diameter first stage, the same as the earlier missiles. This first stage boosts the rocket into the stratosphere, where it burns out and falls away. The fifty-two-inch diameter second stage was from Aerojet Corporation. It takes over, thrusting the rocket higher and faster until it too burns out and falls away. The new fifty-two-inch diameter third stage was under development at Hercules Powder Company. It will continue to power the remaining fourth stage and warheads toward the target. The fifty-two-inch diameter liquid fueled fourth stage was under development at Bell Aircraft Corporation. In other words, the whole rocket was an eclectic mix—what in the programming world would be referred to as a "kluge."

The same could be said of the Apollo program. The first stage of the Saturn V booster was being built by Boeing in New Orleans. The second stage was under construction by our parent company, North American Aviation, at our Seal Beach facility not far from Autonetics. The third stage was being built by Douglas Aircraft Company, in Huntington Beach. Adding to that, North American was building the command and service module. In both cases, the projects were too big, too complicated, and too difficult to be accomplished by a single company.

* * *

The clock was ticking. A lot of work had to be done to write a computer program to simulate the flight of Minuteman III. The first thing was to understand the rocketry. A missile is only a rocket with a bomb on top. In this case, up to three bombs. *How difficult could it be?*

I began to understand the rocket's design. The first three stages have solid propellants. There are four moveable de Laval nozzles at the bottom of the first stage combustion chamber arranged in a cross configuration. I visualized the nozzles being on the arms, or axes, of a plus (+) sign, about five feet across. One nozzle is located on the upper axis of the cross, and one on the lower. Similarly, there are nozzles on the left and right axes of the cross. Each nozzle can be rotated back and forth up to eight degrees around its axis using a hydraulic actuator. The rotation of a nozzle causes the exhaust flame emerging from it to follow, changing the direction of its thrust.

If the pair of nozzles across from one another on the vertical axes rotate together in the same direction, their thrust forces tilt off the centerline of the rocket that, in turn, creates a torque on the rocket to start it turning. The same thing applies to the pair of nozzles on the horizontal axes. Therefore, using the four nozzles, the rocket can be steered in any direction.

Moreover, if the nozzles across from each other are rotated in opposite directions, the rocket would start rotating around its centerline. I grasped the elegance of this design right away. Like in an airplane, the three motion directions are given by the terms "pitch," "yaw," and "roll." Control of all three of these motions is referred to as "attitude" control.

Each of the second and third stages of the rocket have single nozzles that cannot turn to steer their exhaust flames. Instead, a novel method is used to deflect the thrust force in two different directions: there are four sets of small openings spaced ninety degrees apart in the exit cones of each nozzle through which a cold liquid is injected: Freon in the second stage, and strontium perchlorate in the third stage. The liquid vaporizes almost instantly, creating a shock wave disturbance in the exhaust slightly altering the direction of its thrust force, thereby allowing steering in pitch and yaw.

This liquid injectant thrust control method, however, cannot control the roll axis of motion. So, to provide roll control for the second stage, there are two pairs of small thrusters pointed sideways which can be rapidly turned on and off. These thrusters are fed from a small gas generator.

The third stage is similar to the second stage but has only a single pair of thrusters. Firing these thrusters not only imparts a roll to the rocket, but also a slight yaw or pitch motion.

It was exciting to be working with a real rocket now! None of my amateur rockets had the ability to control their attitude. They all used simple fixed fins to help guide the rockets into the sky.

Soon, I focused my studies on the design of the new fourth stage, which was called the Post-Boost Propulsion System (PBPS) — the same diameter as the third stage, but only a bit more than two feet long. It utilizes hypergolic liquid propellants. The fuel is monomethyl-hydrazine, and the oxidizer is dinitrogen tetroxide. The term "hypergolic" refers to the fact that the propellants spontaneously ignite when they come into contact with each other, enabling the engine to start and stop rapidly as needed.

The single fourth stage Rocketdyne engine is gimballed around the central axis of the stage, allowing it to rotate in two different directions, directing its exhaust to affect pitch and yaw motion. Additionally, the PBPS has ten small thrusters, using the same propellants, arranged around its periphery to provide fine control of attitude. Writing a computer program to simulate the solid propellant stages of Minuteman III was one thing, but simulating its PBPS would be quite another.

I recalled the Atlas vernier engine Ron Gallop had serendipitously "borrowed" from Production Machine over six years before, and thoughts of building our own liquid fueled rockets. In the Atlas ICBM, these vernier engines, similar to the PBPS axial engine, one on each side of the Atlas, were also gimballed for steering.

* * *

"So, you're going to write a simulator for Minuteman III." It was Al Sheue, who noticed I had been poring over the documents on my desk for the past few days. His tone was not quite a question, but more like a dare, a throwing down of a gauntlet.

"Yes, I am definitely willing to give it a try," I said. I would try to simulate a missile that, except for the first two stages, did not yet exist.

"Good luck for that..." he said, as he turned and walked away. Then he stopped, and turning back to me said, "Come on, I'll buy you a cup of coffee." Al had saved my bacon previously with the double buffering idea. Now, I was hoping he could be of help in building my new simulator.

* * *

All four of Minuteman III's stages are controlled by a small computer riding in a section called the Post-Boost Control System (PBCS), mounted above the PBPS. Signals are transmitted to/from the various stages along a cable raceway running down the side of the missile.

The PBCS is a bit shorter than the PBPS, and part of it protrudes down into the space occupied by the PBPS to help reduce the total length of the missile.

145

This is important since Minuteman III was intended to replace the existing Minuteman missiles without needing to modify their silos.

The PBCS performs guidance and control. Somehow, in some mysterious way, it is able to cause the missile to find its way from its launch point to a target six thousand miles away with bull's eye accuracy. I came to realize that G and C constituted the true heart and brain of the system, and all the rocketry only served supporting roles. This would take me the next year and a half to truly understand.

The *heart* of the PBCS is the inertial measuring unit (IMU), a much smaller version of the SINS I had dealt with during my work on at-sea gravity measurements. The IMU has a platform gimballed in three axes and stabilized by a pair of spinning gyroscopes. The angular position of each IMU axis is measured by an extremely precise "resolver." The resolvers use inductive methods to measure angles between a pair of hinged mechanical structures that are free to rotate. In the IMU, they are mounted between the joints of the ring-shaped gimbal structures and use alternating current (ac) excitation to make analog measurements that are subsequently converted to digital values for the onboard computer.

Electromagnetic "torquers" on the gimbal axes are used to apply small torques to align or re-align the stable platform as needed. Each of the circular gimbals are supported by a pair of bearings opposite to one another. The resolvers are mounted on one side and the torquers are mounted on the opposite side. The torquers are simplified electric motors that utilize electromagnetic interaction between a fixed and a rotatable part to provide torques between the two parts. The magnitude and sign of the torque is determined by the current in the motor coil which is controlled by the onboard computer.

Three extremely sensitive pendulous integrating gyro accelerometers (PIGAs) are mounted on the stable platform. The PIGAs measure accelerations and, by their physics, accumulate the accelerations into velocities, or speeds, in any direction. Also included as part of the IMU are several sensitive optical devices for aligning the platform when the missile is in its silo.

The *brain* of the PBCS is the onboard D37D computer that runs the flight program, making everything happen. There is a box of electronics for operating the gyros, PIGAs, torquers, and resolvers of the IMU. This box also communicates with the D37D. A second box containing electronic amplifiers interfaces signals from the D37D to the actuators on the missile, for example, the hydraulic steering units for the first stage nozzles. The PBCS also has a battery with enough charge to keep everything running during the short flight time of the missile. I thought it would suffice for me to simulate only the functions of the electronics, leaving out their internal details.

The payload mounting platform sits on top of the PBCS. It incorporates penetration aids, including dispensers for releasing clouds of fine metal wires (chaff), and decoy re-entry vehicles. Up to three Mark 12 re-entry vehicles (RVs), built by General Electric, can be carried—each containing a W62 thermonuclear warhead with a 170-kiloton yield. An RV has the shape of a sharp cone, about six feet tall and twenty-one inches wide at its base, weighing between seven hundred and eight hundred lbs. A protective nose cone, or shroud, completes the top of the missile during its ascent through the atmosphere.

The Minuteman III post boost vehicle, PBV, comprising the PBPS, PBCS, and payload can maneuver in space to deliver its RVs and penetration aids, one by one, toward up to three narrowly-spaced targets—the world's first implementation of MIRV.

I began to think of Minuteman III as a robot, not unlike the robot Phil and I had built using feelers for sensing the room's walls, and telephone relays for a brain. Our little robot's singular goal was to find its way out of a room—sensing, thinking, acting. Minuteman III has its IMU for sensing, and D37D computer for a brain. Its singular goal is to deliver the RVs to the targets—sensing, thinking, acting. A robot.

Lying awake in my bed, unable to sleep, it was strange to think of my own *heart* and *brain*. Holding my breath, I could feel my heart beating in response to electrical signals, pumping blood through my body; perhaps analogous to the electromechanical IMU. My heart, the traditional ancient home of emotions, was lonely for companionship. My brain, with its myriad of electrical signals, was a bit like the D37D. It had to ignore my heart to stay focused on the monumental task that lay ahead. Best to focus on technical things to stop from thinking too much about Minuteman's horrendous destructive power.

19 The Junk

B Y SPRING, 1967, I had started doing most of my own cooking to save money. I packed my lunch to bring to work: ham sandwiches or fried chicken, a banana, and Hostess cupcakes. I would usually get a box of milk or orange juice from the garbage truck.

Gareth Chang and Dennis Nevin worked in my building. We had run across each other a few times but hadn't talked much. They both struck me as happy-go-lucky characters who, unlike me, actually had lives outside of work.

"So, you're from Kansas?" asked Dennis.

"Yes, from Wichita."

"Have you ever been sailing on the ocean?"

"No, I've never even been on a boat before."

Gareth piped up, "Dennis and I own a Chinese junk. We plan to sail it from Redondo Beach to Corona Del Mar. It's only about 70 miles." Then, he asked, "Do you want to come along?" I couldn't really spare the time, given how busy I was with my program to simulate the flight of the Minuteman III, but it sounded like it could be an exciting, daring adventure. "Okay," I said, without thinking about it too much. I figured if they owned a boat, they must know how to sail it.

We drove west toward Long Beach early in the morning of Saturday, April 29, 1967. We stopped at the house of a friend of theirs named Charles, evidently some kind of genius they seemed to respect greatly. Charles was building a submarine to cruise around a mile beneath the surface of the sea. He had already spent ten thousand dollars of his own money on the project. My reaction was skeptical to say the least. I thought of Captain Nemo and *Twenty Thousand Leagues under the Sea.*

"Do you want to see the submarine?" asked Charles.

"Sure," I replied. We walked around the house and into the back yard. There it was, an enormous steel cylinder twenty-five feet long and five feet across. Twin screws and steering vanes projected from the stern, and a mechanical arm extended from the bow attached above a small round window. A stubby cylindrical hatch and periscope pierced the top, and a pair of twenty-foot-long ballast tanks ran along the bottom.

Charles and I climbed the ladder, opened the hatch, and went down inside. There, I was treated to a plumber's nightmare of pipes, air tanks, valves, and carbon dioxide purification systems.

"We're planning to test it this summer," he said. "The Scripps Institute of Oceanography is interested in it and may buy it from me for one- or two-hundred thousand dollars, after it's all checked out." Charles was an amazingly talented person. I was terrifically impressed with what he had already achieved.

Charles drove the four of us to Redondo Beach. We arrived at King Marina around ten o'clock in the morning. Dennis' and Gareth's thirty-foot-long junk was there, tied up in a slip. It looked old, like something I may have seen once in a movie.

We climbed aboard to check it out. Inside the cabin were paintings of oriental gardens and Chinese people. Much of the boat was made of bamboo. "It was built in Hong Kong, and has a gasoline engine," said Gareth, adding, "There's five thousand pounds of lead in the bottom for stability."

It was cloudy and windy. The three of us started out at noon, while Charles stayed behind to send us off. As we motored out to the harbor entrance, frightful waves were coming over the breakwater. I hung on to the forward mast for dear life as the ship battled against ten-foot breaking seas. We were taking water over the bow and stern as the boat pitched and rolled.

Dennis yelled, "We're not going to make it out. The motor isn't strong enough!" It couldn't push the ten-thousand-pound boat into the wind, so we were not making headway. He called out to Gareth, "I think we can try to sail out! Raise the mains'l and keep it about twenty degrees off the wind!"

We made it past the entrance into the open sea. The feeling was exhilarating as I hung on, flexing my legs and focusing my gaze on the horizon. But after a few moments a loud popping sound startled us as our sail was ripped to shreds. The wind was pushing us back against an enormous pile of jagged rocks. After a breathless few moments, Gareth and Dennis got everything under control and we made it back to the calmer waters of the harbor. Happily, Charles was still there waiting for us. This brief misadventure was a lot more exciting than tubing with Ron Gallop on the Ninnescah River.

Spare sail cloth was stowed in the junk, so we went out in the parking lot and spread it on the pavement. Gareth had a needle, thread, and scissors, so we began sewing it all together. We worked from four o'clock in the afternoon to two o'clock the next morning in the freezing wind, illuminated by Charles' headlights. Our fingers were numb as we sewed in strips of bamboo and ropes from the remnants of the old sail.

"These are called battens," said Gareth, referring to the bamboo strips. "The sail is supposed to be sewn together with cat gut using special secret knots." Looking up at me, he asked with a grin, "Happen to have any cat gut on hand?" I got the impression that Gareth knew something about sailing, but Dennis struck me as somewhat lacking in that department. I, of course, was what they call a "landlubber." We made do, hoping our new sail would hold together with heavy cotton thread and granny knots.

That night we slept aboard the junk, nearly freezing to death. I couldn't stand the cold, waking up every fifteen minutes. Charles and I got out and slept the rest of the night in his car.

It was a clear, bright morning; the wind had died down a bit from the day before. Dennis made some fresh coffee while we finished and installed the new sail. Once again, we motored out as Charles stood on the dock and waved us goodbye. Despite the reduced wind speed, the sea was still pretty rough as we reached open water.

Gareth called out, "Hoist the mains'l!" Dennis pulled on the halyard and made it fast. I stood on the foredeck, holding on to the foremast as the ship bobbed and weaved. People on shore were surprised at a Chinese junk with its bright red sail. They stared at us and started taking pictures. I was having a great time as we continued for an hour, tacking our way out to sea where bigger boats were plying. Everyone waved at us as we set a course southward toward Corona Del Mar.

When we turned south, the carabiner connecting the top of our sail to the mast suddenly broke free, and the entire sail came crashing down. The boom, held at an angle from the horizontal, came down hard—smashing Dennis' hand against the cabin roof. He let out with a scream followed by, "My fingers are numb. Something might be broken."

Gareth climbed the mast, fixing the connection with loops of wire, then twisting them tightly. Even though he was only twenty feet above the deck, his bravery impressed me. Once again, Dennis hoisted the sail, but salt water had gotten into his skinned knuckles making it hard for him to work.

The sail held. After a few hours we were getting hungry, so we ate some old bread and chunks of moldy cheese. Dennis broke out a can of pop and poured it on his knuckles. It was then we discovered the kitchen was four inches deep in water and all kinds of moldy white ick was floating around in the fresh water supply tank.

It was mid-afternoon when we spotted the point off Palos Verdes peninsula. Soon thereafter, the wind dropped to zero although the seas continued to be rough. Our nice red sail hung limp—and we had fifty miles to go.

We were still pitching pretty badly when we decided to start the engine to make headway under power. An hour later, the exhaust pipe blew apart and the engine stopped. "Why did it stop?" I asked.

"Couldn't hack the reduced back pressure," said Dennis, as we sat there bobbing aimlessly. By now, the kitchen was eight inches deep in sloshing sea water, and some of the floor panels in the bedroom were starting to float around.

Dennis fished around in the engine compartment for hose clamps and grabbed some pop from the kitchen. Using a can opener and a sharp knife, he fashioned a tube from one of the cans. I helped him slip it over the ends of the disconnected exhaust pipes, crimping it down with the clamps and tightening them with a screwdriver.

We re-started the engine, and the pop-can fix seemed to work. We turned our attention to the water problem. The boat had an electric bilge pump, but we were afraid to use it since the generator connected to the engine was burned out and using it would run down the battery which we would need for the running lights if we were out until dark.

Dennis went below and began the arduous job of pumping the handle of the manual bilge pump up and down to reduce the water level. Eventually, the leaking exhaust fumes overcame him and he started vomiting. He had to come up for air, but up on deck his condition worsened as his injured hand and seasickness took over his senses.

"What's that smell?" Gareth noticed another odor floating above the vomit. With only two of us now taking shifts to man the bilge pump, the water had reached the drive belts in the engine room, and they had started slipping and steaming with the smell of burnt rubber.

It was dusk, and we were all getting cold and tired. Suddenly, Gareth yelled "Turn on the running lights!"

I came up from below in time to see an enormous black wall of steel moving at a high clip off our starboard side. On its port bow was the name BRASILIA, in large white letters. What a thrill! I thought it might be on its way to Brazil through the Panama Canal after dropping off a load of bananas.

"That was a close call," said Gareth, grimly. "They didn't even see us." Listening to Gareth's pronouncement, it crossed my mind that we might all die. The junk creaked and groaned as it began to roll from side to side in the wake of the ship. "I'm not feeling so good." Gareth was beginning to succumb.

Before the darkness of night descended, we got a visual fix on the farthest thing away, which we guessed was some structure in Huntington Beach. After hours of motoring in that direction, we discovered it was not Huntington Beach after all—but a huge drilling platform far out in the ocean. Gareth managed to get some shots of the platform tower lights on his Polaroid. They gave us a loud blast on their horn to warn us away. We didn't want to get close enough for them to start shooting.

Our running lights had been turned off again, since we didn't want to run the battery down. Without a battery and generator, the engine would stop running, leaving us bobbing around as a powerless speck in the blackness of the ocean.

The exhaust system blew out again. Apparently, the hot exhaust gas had eaten its way through the thin aluminum pop can. This time, miraculously, the engine more or less continued to run. We were glad of that, since it was suicide to go down in a dark hole to fix it again and man the bilge pump while standing in knee-deep sloshing water breathing exhaust fumes.

Gareth pulled out three life jackets from a compartment in the ceiling below deck. "Here, put these on now and make sure the straps are tight." As I put on the World War II era kapok life jacket, I started thinking about the five thousand pounds of lead ballast beneath my feet. We wanted to be topside so we could jump, and not be sucked under when the whole thing went down.

Meanwhile, we were trying to plot a course using our loop antenna radio direction finder—harking back years ago with Ron Gallop's transmitter hunts. As we slowly rotated the loop, listening for the highest signal, we tried to find the strong Boss Jocks radio station. But, to our surprise, we found our navigation chart was printed in 1936. All they had on the chart was Rudy Vallee's and Tom Mix's stations. We couldn't find a peak signal among the plethora of stations. It didn't matter, since we couldn't draw a straight line in the rocking boat anyway.

Darkness now was complete, but the sky was filled with stars. A few lights were visible on the horizon; we couldn't tell if they were on shore, ships, or drilling platforms. After much discussion between the two sailors on the principles of nautical navigation, Gareth said, "I think we should continue in a generally southerly direction." He was pretty sick by now, and had vomited several times. Curiously, I had yet to feel seasick.

Hours later, freezing and cold, we had to change course since we were beginning to make out a gigantic rock pile, dead ahead of us. After making our way around the rock pile, the entire coastline of southern California abruptly appeared with millions of lights. It had been hidden behind the rock pile. "Can you see Corona?" Dennis asked Gareth.

"I'm trying to make it out, but all the lights look exactly alike." We spent some long hours trying to solve the problem.

Then it happened. A line that had fallen overboard earlier, perhaps when the sail fell, and had gone unnoticed up to now—snapped taut and became fouled in the propeller. Dennis cut it loose with his knife, but the engine started coughing, reducing our speed to only a few knots. Nevertheless, we continued to make some headway.

"I think we should just head for the lights and beach it," said Dennis. "Maybe someone can rescue us." He and Gareth were sick, cold, and tired.

"Most of the shoreline is rocky. If we tried that, we would break up on the rocks and have to swim for it," said Gareth.

"I'm a good swimmer, but I don't think I would make it in the cold and waves," I said. We knew we were riding lower in the water. I grabbed the flashlight, and shined it down into the cabin, trying to make out the water level in the darkness. "Hard to say, but well over knee deep," I announced. "I think we should continue heading toward the lights, and when we get close, we can go south along the coast toward the harbor."

"That will take too long. By that time, we'll be goners," replied Gareth.

A thrill of excitement was starting to well up in me. I thought perhaps we could navigate using the stars. I soon found my favorite, the North Star, Polaris. I could fixate on it as the boat weaved along. Looking along the distant horizon, I saw that one of the brighter lights was blue. It occurred to me that the blue light could possibly be a giant sign on top of the Union Bank building in Santa Ana, ten miles inland. I had driven by it countless times on the Santa Ana Freeway. "Hand me your flashlight again," I said to Dennis. The navigation chart lay flat on my lap. The town of Santa Ana was marked on the map, along with a few other towns in Orange County. In 1936, the towns were discrete places surrounded by orange groves and strawberry fields, not yet merged together to form a continuous megalopolis. Corona Del Mar was marked on the map as well, directly south of Santa Ana. I had the solution. "I know how to bring us into the harbor."

"Okay, why don't you tell us?" asked Gareth.

"All we need to do is continue sailing until the blue light you see in the distance is directly under the North Star. Then we can turn and head directly to Corona. I think the blue light is a sign on top of the Union Bank building in Santa Ana."

"Let me see that," said Gareth. He too, understood the idea. "Okay, right now, the blue light is to the right of Polaris. If we continue, we should see it move in line. I hope you're right about the Union Bank sign."

It appeared to be the solution we were seeking. As we made our way southward, the blue light moved closer in line with Polaris. When the light was directly beneath the star we turned left, heading straight north. We continued to steer the junk to maintain that relationship.

"We need to man the bilge pump in short shifts. Hold your breath, go down and pump, and head back up," commanded Gareth. He and I did that. Dennis tried his best, but with his injured hand and seasickness he couldn't manage more than two or three strokes on the handle.

Two more hours passed by. With the boat riding ever lower in the water, the pall of dread hanging over our waterlogged vessel began to lift: we spotted the flashing beacon light of the harbor, right on target. "That's it!" said Dennis. In another hour, we made our way through the darkness amidst clanging buoys and past the breakwater to the dock, safe at last! Gareth climbed out and tied off lines to the bow and stern. Then Dennis made his way up. Finally, it was my turn. Gareth ran a power cord to a box on the dock, starting the electric bilge pump. The three of us stood there in the darkness—cold, wet, and shivering while waiting for water to start exiting the boat; tight-lipped glances between us revealing both our agony and relief. None of us uttered a word. Climbing the stairs to their apartment, I collapsed on the sofa and instantly fell asleep.

20 A Computer for a New Robot

THE FIVE THOUSAND POUND wrecking ball came crashing down from the crane's mast—obliterating the big beautiful blue and gray IBM 1620 computer with its twenty thousand decimal digit core memory. A thousand more 1620s awaited their fate in the yard of the vast surplus electronics store near the Oakland, California airport.

I didn't actually witness this happening—I read about it in a printed flyer. IBM had contracted with the surplus store to methodically destroy the 1620s, preventing them from competing with their next generations of computers. An inspector from IBM periodically audited the progress of this destruction.

Apparently, the crane operator eventually learned to hit the 1620s with such an artful blow that the core memory stack with its twelve layers of tiny ferrite donuts, ten thousand donuts to a layer, and thousands of hair-thin wires snaking through the whole thing was spared. Some enterprising hobbyists from San Diego had purchased a few of the priceless stacks for a song. I immediately realized I needed to get hold of one of these memory stacks for the computer I had been building on my kitchen table.

Since Phil and I had built the small mobile robot using telephone relays for a brain, it became obvious that having a real digital computer for a brain would give the robot a lot more intelligence. But computers were far too big and heavy to mount onboard the robot. There would need to be some sort of radio connection between the two. If I was rich I probably could have bought a computer like the Digital Equipment Corporation's new PDP-8 that had a price tag of over $18,000, more than twice my annual salary from Autonetics.

Being nowhere near rich, in 1967 I started building my own computer. Phil and I had begun building a new robot when I was a grad student at Kansas State. But since then, the project had languished for lack of resources and the fact that Phil was working so hard to graduate from Caltech with his Ph.D.

I had already etched a large circuit board in my bathtub, drilling all the holes on the Sears drill press I had installed between my stove and refrigerator. "What's causing all the noise in there?" asked the landlady at my door. "I'm getting complaints from the other renters."

"It's just a malt mixer. I'll try to be quieter," I said.

I had started wiring up logic integrated circuits procured from the local Texas Instruments office. They were working with Autonetics on building Minuteman III D37D flight computers. When I told the TI people I was building my own computer, they were happy to give me some of the parts I needed.

I had talked with engineers at the local office of DEC who gave me a bunch of paperback-sized manuals covering the PDP-8 and their extensive line of plug-in circuit boards. I spent many hours soaking up the contents of these small books. But the one thing I needed most—a core memory—was unavailable. Until now!

Rocky and I boarded a plane at LAX and two hours later we were at the surplus store in Oakland. The man there told us that no more core stacks were available. He told us the IBM guy had caught wind of their scheme and now had forced them to take the top off each 1620, remove the top cover of its core stack, and plunge a claw hammer down through all the layers of little ferrite donuts, destroying everything.

Shocked and dejected, we were ready to leave, when he said, "Would you be interested in getting some individual core planes? We found that plunging the claw hammer only part way down seemed to satisfy the IBM guy. Anyway, there was no way he could see through the maze of wires all the way to the bottom. So, we've been able to salvage two or three planes from the bottom of each stack."

The sixteen-bit word length of my computer design needed two extra flag bits. "I'll take eighteen planes," I said. Instead of the 1620's twelve planes encoding twenty thousand decimal digits, my computer would have ten thousand sixteen-bit binary words.

Back in my apartment, after midnight, the fifty-thousand-watt XERB border blaster was coming in strong on my radio.

"Heh heh. Talkin' about getting' blown away... Old grandpa Cretin beginnin' to slip in the memory department. Yah, bein' a hundred and seven is really affecting his mind. Jus' the other day he accidently threw a man out of the thirteenth story window. When the police asked him why he did it, he he told them he used to live on the ground floor but he just forgot, heh heh. Guy wasn't hurt though, he just kinda shook up his legs a little..."

Listening to the Wolfman's gravelly voice for entertainment powered me through un-soldering the seven-thousand two-hundred bits of wire from the edges of the planes, stacking them up eighteen high, and over a week of nights soldering the necessary ten thousand new wire connections for my computer.

XERB had descended from XER, across the border from Del Rio, built by Dr. John R. Brinkley of goat-gland and patent medicine fame. As a college student in the 1930s, my father had once played a clarinet solo on KFKB, Brinkley's pioneering station in Milford, Kansas.

The finished core stack needed drivers to send currents through its addressing wires flipping the magnetization of the ferrite donuts from clockwise, representing a "zero," to counterclockwise, representing a "one." Phil and I spent an entire Thanksgiving at his house wiring up an array of larger donuts to act as drivers.

21 Minuteman Flight Simulator

I NEEDED somewhere to start designing and building the simulator. By now, my FORTRAN programming skills were pretty good. But shortly after I embarked on the project, Autonetics took away the 7094 computer. At this point, I had not done a whole lot, but the departure of this great computer was certainly consistent with the story of my life. It seemed that once again, just as I was getting good at something, it was taken away.

In its place, a strange brand-new computer, the IBM System 360, was being installed in the Data Center. The 360 had the capability to pretend it was a 7094. This so-called "emulation" mode made it possible to run old 7094 programs on the new computer.

Nevertheless, it was decided that my new program, that I started calling the *Minuteman Flight Simulator (MFS)*, would not use the 7094 emulator, but instead would be programmed in an updated FORTRAN language, FORTRAN IV, using the native System 360 instructions.

A word in the System 360 has only thirty-two binary digits (bits), as opposed to the 7094's thirty-six bits. People around me started talking about "bytes." I didn't have a clue what a byte was. Soon, I learned there were eight bits in a byte, and four bytes in a System 360 word. Each byte could be expressed as a hexadecimal number, 0-9 and A-F, pretty much like in the Bendix G-15, but using the letters A, B, C, D, E, and F, instead of u, v, w, x, y, and z. By now, I could do octal arithmetic in my sleep, so this was a big change to wrap my head around.

Another important consideration concerned me. With only thirty-two-bit words rather than thirty-six, there would be a big reduction in precision. Accordingly, I decided to use double precision arithmetic, sixty-four bits, in all the calculations to be performed in MFS, giving roughly fifteen decimal digits of computing accuracy.

I needed to simulate, as closely as possible, the Minuteman III missile and its environment during flight. This would enable Autonetics to check out and validate D37D flight programs used to fly real missiles. As such, it would be used to demonstrate the flight program's ability to navigate and steer the missile through powered flight as programmed, perform correct attitude maneuvers and object deployments, and provide correct commands in the proper sequence to the missile subsystems.

Over a few weeks time I came to realize that major areas of the simulation would have to include the D37D computer, the inertial measuring unit (IMU), the vehicle structure including staging events, nozzle movements, the effect of digital output signals, the atmosphere, the aerodynamic properties of the missile's airframe, and the Earth's gravitational field.

There would be two main inputs to the simulation: data describing the missile and its environment, and the D37D flight program itself along with its targeting information. The simulation would output printed and graphical results showing the time history of the simulated flight. The MFS program would also need to have provisions for users to specify control of its operation. Considering the huge challenge that this represented, I began to experience a deep fear in my bones. It may be much more than I had bargained for.

* * *

I learned that the D37D computer inside the PBCS is about the size of a large shoe box, containing circuit boards with several thousand radiation-hardened integrated circuits (ICs) made by Texas Instruments. Its main memory is a rotating magnetic disk on which programs and data reside. The disk is the moral equivalent of the magnetic drum memories I had experience with for the IBM 610 and Bendix G-15. The disk rotates at six-thousand revolutions per minute, supported by a frictionless gas bearing. Multiple circular tracks, or channels, are written on both sides of the disk, with each track divided into a large number of radial segments, or sectors. The D37D doubled the size of the Minuteman II D37C computer memory to 14,137 words and increased the number of discrete, or individual, inputs and outputs to 110 and 74 respectively. A D37D word contains twenty-four bits.

The D37D disk has fixed read/write heads to access all of the channels. Several special channels are set aside for high-speed access. Heads in the main special channels, designated H, V, R, D, and U, read or write data from/to integrated circuit memory, and immediately write or re-write the data a short distance "upstream" on the channel so it can be read or written again without waiting for the disk to make a complete revolution. Because of this action, these channels are referred to as "loops." Thus, for these special channels, there are multiple copies of the same data around the disk, enabling rapid access.

The V-loop reads velocity information from the stable platform PIGAs. The flight program uses this information to perform guidance calculations to continually update the vehicle's velocity and position in space. The R-loop reads precise angle information from the resolvers on the IMU's stable platform gimbals. The flight program, using its navigation calculations, combines this angle information with the desired vehicle attitude, writing results of the calculation into the H-loop to provide control for the vehicle's thrusters. The D-loop controls the ten small attitude thrusters on the Post-Boost Propulsion System (PBPS). The U-loop is used to compute the precise fine-countdown time for thrust termination or object deployment.

Simulating the operation of the D37D is a daunting task. Fortunately, Robert Best in the Data Systems Division was already busy writing a program called DSIM to simulate the Minuteman III D37D flight computer, modifying earlier versions to run on System 360 using the 360's native assembler language.

DSIM is interpretive, simulating the operations the D37D performs and the effects of these operations on registers and memory. A large area of the System 360 memory is set aside to accommodate the D37D disk memory. This "pseudo-memory" contains all the channels and sectors, including all the rapid access loops. It also keeps track of their respective timings. Areas of System 360 memory are also provided for the D37D program counter register, accumulator register, and several other special functions.

DSIM encodes a twenty-four-bit D37D word inside a System 360 thirty-two bit word. As such, it could be considered to contain three eight-bit bytes, but for historical reasons is treated as eight 3-bit octal characters. Once again, I was right back to thinking in octal. All of this continued to be very confusing.

At the beginning of simulation, the flight program and target tape are read into the DSIM pseudo-memory. Execution of these pseudo-instructions then begins at some specified location. For example: let's say the first instruction to be executed is the instruction "104.125 ADD 02.136, 147." This instruction resides in channel 104, sector 125. It says to add the contents of channel 2, sector 136 to the accumulator (A register), and take the next instruction from channel

104, sector 147. Inside the pseudo-memory, this 24-bit instruction is coded in octal format as "65470336."

DSIM decodes this octal pattern into the operation code, next instruction sector, and operand address. A branch is taken to a routine that simulates the "ADD" instruction, thereby altering the pseudo-accumulator. The pseudo-instruction location counter is then updated, and the next instruction is taken. The entire flight program is executed in this instruction-by-instruction fashion.

Sy Alper's one-of-a-kind analog computer simulation had the luxury of being hooked up to a real D37D, so it could operate in real time. In other words, his analog computer could output results at a speed to keep pace with the flight of the real missile.

It was easy for me to grasp that an all-digital simulator of Minuteman III would have to run much slower than the real missile. But, on the other hand, it could be much more accurate, output many nuanced details of flight events, and could run on any IBM System 360 computer.

DSIM would serve as the main program of MFS, and I would need to write a bunch of subprograms that would interact with DSIM. Like the D37D computer controls the operation of the real physical missile, DSIM will control the operation of my imaginary simulated missile. This is the mission I dived into. Somehow, I needed to craft my way into the heart and brain of the beast.

* * *

I understood that MFS had to faithfully execute whatever the flight program told it to do and return the status of the missile back to the flight program so it could do whatever is necessary to control the missile. But who were the people tasked with writing the intricate flight program code to operate the missiles that would defend America?

A half-dozen of them were housed in a nearby cubicle: John Hlavac, Gary Wilson, Bruce Simpson, Hope Cope, Maryann Carlson, and a few others. They were all going crazy trying to deal with the idiosyncrasies of the D37D, or it's possible they were all crazy to begin with. Every now and then they would march around, shout, and throw waste baskets full of punched cards and listings over the cubicle walls, inundating the occupants of neighboring cubicles. John Sweet, their manager, rode herd over them.

John Hlavac exemplified the "Flight Weenies," his name for the group of flight programmers. He was short, balding, asthmatic, and hilariously funny. We soon became friends. His taller alter ego was Alan Bernstein, who worked on the Minuteman III ground program; these two were inseparable. Alan was a well-dressed east-coast Jew who effused a barely detectable odor of naphthalene stemming from the moth balls standing guard over his fine wool wardrobe.

Hlavac and Bernstein were a comic duo in the tradition of Laurel and Hardy, or Abbott and Costello. I sometimes had dinner with them—and they kept me in stitches with their rapid back-and-forth bantering. Hlavac knew exactly what to say in response to Bernstein, and vice versa. Both of them were cynical and defeatist when it came to writing programs for a computer that had a rotating disk for its main memory.

The comic duo was trying to connect with women. They took pity on me and somehow found three girls to smooch with us in John's darkened apartment. A large girl was assigned to me. She sat on my lap, and we hugged for a while. That was about it; nothing more exciting to report.

Another guy at Autonetics, Richard, was studying part-time to become a doctor. He approached me at my desk one day. "Ralph, we're having a Wesson Oil party at my apartment next Saturday night. Why don't you come? It's a lot of fun."

I had no idea what he was talking about. "No, I don't think so," I replied, too much of a prude to inquire about the details.

* * *

"John, how many lines of flight program code can you write in a year?" We had a chance to chat a bit on more serious matters, and I wanted to try to understand how the flight program worked. I had walked the short distance from my cubicle to Hlavic's.

He looked up from the mess of paper littering his desk. "Oh, maybe about fifteen lines in a year. The D37D only has about fourteen thousand words, and half is off limits since it is used by the ground program. It's hard as hell to cram in enough code to do guidance and control with only the remaining seven thousand twenty-four-bit words to play with. At this point we're spending day and night trying to save a few dozen words."

"Don't you have any kind of assembler or compiler?" I asked.

"We have Jerry Brown's Merlin assembler, but it doesn't help much because of all the timing constraints of programming a rotating disk main memory. As I'm sure you realize, if an operation takes too long to execute, it will miss the next instruction in sequence, so we have to wait until it comes around again on the disk. I'm talking about the horrible slipped-disk problem; throws off all the timing, so we crash. Will MFS check for slipped disk events?"

"Well, I guess so," I said, hopefully.

"Then there's semi-somnus," he said.

"What's that?"

"They think the Russians are beginning to test high altitude bursts creating X-rays that can fry our missiles as they leave the silos. Sensors in the missile's skin detect X-rays. When that happens, the D37D logic shuts down in mid-instruction, letting the disk rotate a few revolutions before starting up again. So, for a short time we're flying blind," he said.

"Doesn't it throw off all the timing?" I asked. I already knew that the Minuteman III PBCS electronics had additional shielding over Minuteman II.

"Exactly. We have to cheat by doubling up on the number of PIGA counts until we're caught up. What a headache!" Although I had plenty of my own problems designing the simulator, I was glad I was not a flight programmer. Over time, I gained a lot of respect for Hlavac and the other Flight Weenies.

* * *

I knew right away that I needed to apply algebra, logic, and calculus to construct the innards of MFS. I also needed to find a program structure to make it tractable and understandable. I thought back to being part of the row of high school students punching away in front of Friden calculators. We had used a simple way to add up accelerations and velocities in each increment of time to simulate our rockets' flights. These operations performed what is known as "integration," in the calculus. I knew I needed a much better integration method for MFS. The best person to ask about this was Al Sheue.

"Al, I'm struggling with the problem of integration. What should I choose for the integration method?"

Once again, Al came to the rescue. "The best thing to use for the problem you are trying to solve is Fourth Order Runge-Kutta. It's accurate and works well when the interval between calculation steps varies."

"Do I have to write a FORTRAN program to implement Runge-Kutta?"

"Nah, it's already written. You just call the Runge code, then calculate all the equations of motion and their interactions, and when you're done with that, you call Kutta. It's pretty slow, though, since it has to take four passes through your equations for a single time step."

Carl Runge and Wilhelm Kutta were a pair of German mathematicians who published their integration method in 1901. I found a derivation of the math in a book on numerical methods. It relies on calculating four different slopes of a curve.

After many hours of study, I failed to understand exactly how it worked. I was loathe to include anything in MFS that I did not completely understand, but I went ahead with it anyway. Sy Alper's analog system used vacuum tube circuits called "operational amplifiers" to perform integration.

* * *

166

I usually took a break in the afternoon. To determine who would buy the coffee, we gathered in a vacant office and played a homemade game of *Battleships* for multiple players. Lucille was often the leader, drawing a ten-by-ten grid of squares on the blackboard. The top row was labelled "A" though "J" and the leftmost column was numbered one through ten. The leader secretly filled in their own paper copy of the grid with rectangles of various sizes to represent the different kinds of ships from the famous game.

Eight or ten participants made their own paper copy of the grid to keep track of their moves. To start the game, Jerry, for example, would hem and haw for a while and then call out "E5," after which the leader would mark an "X" in that square on the blackboard, announcing whether it was a hit or miss.

"By the way, Jerry, have you cashed any of your paychecks?" someone would say.

Then, it would go around the room with each of us pretending to be applying some strategy to our selection while making some clever remark that might illicit a laugh. When all of the squares of a ship had been hit, the leader would announce the sinking of the corresponding type of ship. It often took twenty minutes or half an hour to sink all the ships. The player who scored the least number of hits bought the coffee for the leader and everyone else.

Computer programming is a truly exciting endeavor, especially when you can make something work for the first time. But it is often lonely, dull, and stultifying. Sometimes one yearns to go outside and fly a kite or do anything but code up one more routine. Determining who would buy the coffee, even if it might be considered a waste of time, was a vital activity for our little group.

* * *

Frankie and I had been seeing each other for over six months. I called her to ask if she wanted to go for a drive. "I don't think we should see each other anymore," she said. "I have a boyfriend now. We made love last night. He was very gentle."

Her admission hurt me greatly. *I was her boyfriend!* I was quite attracted to her but began to realize—weeks later—we had never hugged, never kissed. Once, we may have held hands. So that's it. I guess I never learned how to date. Most teenagers date around for years before picking a mate, which I never did since I was focused on science and rockets. I got married because my first and only girlfriend got pregnant. My naiveté was astounding. I had to start all over to learn how to find love.

* * *

Three months into designing MFS, I had DSIM and the integration method, Fourth Order Runge-Kutta, under my belt. I had also managed to program several simulations of the rocket stages. I learned that the solid propellant inside each of the rocket motors has a star-shaped opening down their centers, much like what I had calculated for our Cryolite rocket on the Bendix G-15.

In Minuteman, digital signals from the D37D set off igniters at the top of each stage's combustion chambers, causing thrust to develop almost instantly, continuing for about sixty seconds for each of the stages until burnout. The first stage develops thrust of over 200,000 lbs.; the second stage over 60,000 lbs.; third stage over 35,000 lbs. Once a stage is lit, there is no control over it. The actual thrust profile over time is a curve of data points determined from many experiments that were conducted at Thiokol and the other companies. I used a table lookup procedure to model thrust.

Values from the table had to be modified depending on the ambient temperature and air pressure, an additional complication. Of course, as the propellant burns, and protective insulation burns off as it exits the silo, the missile becomes lighter, and its center of gravity moves forward toward the nose. All these effects needed to be taken into account.

I began to grasp the bigger picture of how everything in the rocket is related. At a fundamental level, I knew the missile has both translational motion and rotational motion in space, and each of these motions has three different components. For MFS, I needed to implement a collection of mathematical models, each describing a portion of the problem which would be encapsulated within a subroutine. Most of the models would also have numerous sub-models.

The Translational Motion Model depends on gravity, properties of the vehicle's mass, thrust forces, and aerodynamic forces. The Rotational Motion Model depends on the mass properties, aerodynamic forces, thrust forces, and the "center of pressure." This latter term is an imaginary point along the vehicle's centerline where it is assumed all the aerodynamic forces are acting. All of these effects are continually changing during flight. Like I had reasoned for my little Firestreak rocket design, I knew the center of pressure needed to remain aft of the center of gravity to prevent the vehicle from flipping end for end.

In turn, both the translational and rotational motion affect many other things. For example, my simulated translational motion calculation must produce an acceleration, and by integration must produce the vehicle's velocity, and position. The position, in turn, affects the calculation of gravitational acceleration, and also feeds into the Atmosphere Model, for example, the air pressure, and wind speed. The vehicle's velocity feeds into this model as well, to produce dynamic drag forces.

Meanwhile, the Atmosphere Model must calculate the velocity of the vehicle's airframe relative to the air mass, as well as the vehicle's Mach number, feeding these into the vehicle's Aerodynamic Characteristics Model. The Mach number, honoring Austrian physicist Ernst Mach, is the ratio of speed to the local speed of sound.

In turn, the Aerodynamic Characteristics Model must calculate the vehicle's drag and lift quantities, along with the dynamic pressure from the Atmosphere Model to calculate the aerodynamic forces on the vehicle. It must also calculate the position of the center of pressure used in the rotational motion calculation. However, the aerodynamic characteristics depend on the vehicle's attitude calculated by the Rotational Motion Model. Here, I realized it was a tricky circular situation, like a cat chasing its tail.

There are other looping things happening as well. The Propulsion and Structure Model needs to know the air pressure from the Atmosphere Model to compute thrust values from each stage's rocket motors. During the climb into the sky, as the atmospheric pressure decreases, motor thrusts increase because of the reduced back pressure. It also needs to keep track of the vehicle's mass properties as propellant is rapidly burned away, feeding these into both the Rotational Motion and Translational Motion models, and so forth, like a roomful of cats chasing their tails.

All of these models would also apply as well to our unguided amateur rockets. Of course, Minuteman is more than a thousand times bigger. The only real conceptual difference is the fact that there is a digital computer, the D37D, on board the Minuteman III. The computer inside the missile—without any guidance from the outside—is trying to follow a smooth pre-computed arc-shaped path from the launch point to the target.

This ballistic path is much like the path a giant artillery shell would take. Points along this path, called a "reference trajectory," stored inside the D37D prior to launch, specify the required attitude and velocity of the vehicle at each moment of time. The vehicle's distance along the path at any given time is uncontrollable and will vary depending on the actual thrust produced by the motors. The flight program running inside the D37D strives to follow this reference trajectory. In Minuteman III, after the third stage cutoff, up to three RVs can be released subsequently by the fourth stage PBPS toward up to three closely spaced targets.

I turned my attention to the inertial measuring unit, the *"heart"* of the missile. There would need to be an IMU Model simulating the physics of the gyro-stabilized platform on which the three pendulous integrating gyro accelerometers (PIGAs) are mounted. MFS would need to take angular accelerations computed by the Rotational Motion Model, and accelerations computed by the Translational Motion Model, and feed these quantities into the IMU Model. In turn, the IMU Model would need to write PIGA outputs, called "counts," into the D37D V-loop, and also write stable platform gimbal angle resolver counts into the D37D R-loop.

The V- and R-loops are read by the flight program guidance calculations as rapidly as it can to determine the magnitude and direction of any discrepancy between the vehicle's actual attitude and velocity and the specified attitude and velocity given by the reference trajectory. The discrepancies are referred to as guidance "errors." The flight program then tries to bring the errors down toward zero, using another calculation referred to as a "control law," writing its control efforts into the H-loop.

My program, MFS, must read the H-loop values, representing commanded nozzle angles in the case of the first stage, liquid injector flow rates in case of the second and third stages, and axial engine angles for the fourth stage PBPS, feeding them into yet another model: the Actuator Model. This model must also read data from the D-loop to simulate turning on or off the ten PBPS attitude control thrusters.

The Actuator Model also uses thrust values computed by the Propulsion and Structure Model to compute the direction and magnitude of thrust forces acting on the vehicle. These, in turn, feed into both the Rotational Motion and Translational Motion models. So, in truth, everything depends on everything else. That's the magic of G and C. Could all of this be accurately simulated by a FORTRAN program?

* * *

In the morning of January 27, 1967, my thoughts were suddenly interrupted by a voice on the PA system. "There has been an accident at Cape Kennedy Launch Complex 34. A fire erupted inside the Apollo 1 command module. Virgil Grissom, Ed White, and Roger Chaffee were killed." More details were given.

A great sadness filled the building as everyone packed up and left for home. I had followed Grissom's and White's careers for years. Grissom was one of the original Mercury astronauts. White was the first American to do a spacewalk. The three were performing a "plugs-out" test on the ground, getting ready for the first launch of the Apollo spacecraft into orbit, scheduled for February 21.

Our company had built the command and service modules for Apollo at our Downey facility and was therefore partly to blame for the accident. A complete re-thinking of the design and manufacturing of the command module would be needed to move forward with mankind's attempt to land men on the moon. North American Aviation was in trouble.

Some speculation circulated that North American might be bought by another company. Once, it seemed the Schick razor company would step in to bail us out. Someone in the group joked that Kentucky Fried Chicken should be brought in to help sweeten the deal. Then, the once-proud company could be called "North American Chicken Schick." Two months after the Apollo fire, we were acquired by Rockwell-Standard, a vehicle parts manufacturer based in Pittsburgh. Autonetics' new parent became North American Rockwell.

* * *

I got into the habit of getting up early, putting on my white shirt and black tie, and driving the three miles to Autonetics. After grabbing a bad tasting cup of coffee and a roll from the machines, I was good to go. Usually, by then, half of my area in the bullpen was filled with people.

By now, I was working sixty hours a week, appreciating the overtime pay. Timecards were filled out and turned in on Fridays. All the cards from Autonetics were dumped in a big bag and flown up to San Francisco on the weekends for processing by the federal government, after which a big check was flown back to Anaheim on Monday mornings.

Every month, I signed a form declaring I was doing important secret government work and therefore would not be available for the draft. The war in Vietnam was escalating with something like 300,000 American troops on the ground. Bombing in North Vietnam had begun. I did not want to go there.

The Minuteman Flight Simulator was coming along, but the first Minuteman III test launch was less than a year away. My struggle to understand the full scope of the problem I was trying to solve weighed heavily on my mind. I had long since given up submitting coding forms to the keypunch operators. When a machine was free, I simply punched the cards myself, constructing the program straight out of my head. It was easier at night when there were fewer distractions.

"What the hell are you doing?" It was Dick Snyder. "I'm not paying you to punch cards." His tone was harsh. It was ten o'clock at night, a rare time for the boss to show up.

"I'm used to programming this way. It's easier for me," I said. Mr. Snyder swore something under his breath and walked away. I supposed he was also feeling the pressure of the first launch.

I gave up on special couriers for sending my card deck to the Data Center, opting to carry it there myself. After submitting my job, I would usually hang around until the program ran and failed. This procedure would go on five or six times a day or night, enabling me to incrementally build a working program fragment. Working alone, I experienced the loneliness of code immersion. Sitting in front of a keypunch machine, I tried to will the code into my head; code that would serve to implement a needed mathematical function. Staring into space, thoughts of Frankie would suddenly take over and I would forget what I was doing and would have to start all over.

* * *

I realized that calculations of forces and torques acting on the Minuteman III vehicle would have to be carried out with respect to a specified coordinate system, or coordinate "frame." In most cases, the coordinate frames have three perpendicular axes, denoted by X, Y, and Z, or their Greek letter equivalents. I visualized a frame as the three edges formed by the ceiling and two walls of a room. The origin of this frame would be the point where the edges all intersect.

Days and nights spent going over the mountain of papers on my desk confronted me with nine important frames, abbreviated ECI, ECEF, LCEF, P, C, MBF, MS, G, and PIGA. I needed to understand how these frames were interrelated to design the missile's simulation.

The Earth-Centered Inertial (ECI) frame has its origin at the center of the Earth and is fixed in inertial space. That is, it could be said to be stationary with respect to the "fixed" stars. The X and Y axes lie in the plane of the equator, with the Z axis along the Earth's spin axis extending through the North Pole. As the Earth rotates, the Greenwich, England meridian passes through the fixed X axis at the instant of launch.

The Earth-Centered Earth-Fixed (ECEF) frame has its origin at the center of the Earth, and remains attached to the rotating Earth, but unlike the ECI frame, the X axis remains fixed through the launch site meridian. *So far, so good.*

The Launch-Centered Earth-Fixed (LCEF) frame has its origin at the launch site and is fixed in the Earth like in the ECEF frame. The launch site is specified by its longitude and "geodetic" latitude. If the Earth's shape is represented as an ellipsoid, with flattening at the poles and bulging at the equator due to its spin, the geodetic latitude is defined as the angle between the equatorial plane and the surface normal at a point on the ellipsoid, for example, the launch site. (The surface normal is perpendicular to a plane tangent to a point on the ellipsoid.) The Earth has a rather slight equatorial bulge: it is about twenty-seven miles wider at the equator than pole to pole. The surface normal, or Z axis of the LCEF frame, points upward from the launch site, but because the Earth is not a sphere, the line of the Z axis does not quite emanate from the Earth's center. Furthermore, small deviations in the launch site gravity cause the plumb-bob vertical to differ slightly from the geodetic Z-axis vertical, and this must be taken into account. Because of the local gravity, the LCEF Z axis is considered to be the plumb-bob vertical at the launch site, in other words, pointing directly opposite to the pull of gravity. The LCEF X axis is directed toward the nominal target and lies in the pitch plane of the reference trajectory. This was starting to give me a headache, but I knew it was super important to understand all these coordinate frames.

The Platform (P) frame is ideally the same as the LCEF frame. Since the LCEF frame is used for guidance computations, the flight program must at all times be aware of the orientation of the LCEF frame. The stable platform of the inertial measuring unit (IMU) provides this reference. Prior to launch, when the missile is resting in the silo, torques are continually being applied to the stable platform gimbals. Without these torques, the stable platform would remain fixed in space while the Earth rotated about fifteen degrees per hour. Instead, the torques cause the stable platform to remain stationary with respect to the missile body. During flight, the same rotation rate continues to be applied by the torquers. Assuming the platform is being torqued correctly, and there are no gyroscopic drifts present, then the Platform (P) frame is the same as the LCEF frame. MFS, however, must be able to simulate gimbal torqueing errors and gyro drift, so the P frame will never be exactly the same as the LCEF frame. I wondered how this was going to work.

The Computational (C) frame is another Earth-fixed frame. It is rotated through an angle called Delta alpha about the Z axis with respect to the LCEF frame. Delta alpha is the offset angle between the nominal Platform (P) frame X axis and the actual target direction. Whenever Delta alpha is non-zero, the D37D flight program must correct for the fact that the platform is not aligned with the target.

The Missile-Body-Fixed (MBF) frame has its origin fixed at the vehicle center of mass (CoM). For this frame, some Greek letters come into play. The Ξ (Xi) axis points toward the nose of the vehicle and defines positive roll. (If my right thumb is pointing along the positive Ξ axis, my fingers will curl in the direction of positive roll.) In a similar manner, the H (Eta} axis defines negative pitch. The Z (Zeta) axis lies in the normal pitch plane through the belly of the missile and defines positive yaw motion. The MBF frame is used for the calculations of thrust, aerodynamic, and deployment forces as well as the vehicle angular velocities. All rotational motion calculations are made most conveniently about the center of mass. As the vehicle changes mass during flight, the origin of the MBF frame, the CoM, moves with respect to the component parts of the vehicle. As staging occurs, the CoM abruptly moves in the positive Ξ direction, toward the vehicle's nose.

The Missile Station (MS) frame has axes always remaining parallel to those of the MBF frame. Unlike the MBF frame, however, the MS frame remains fixed in relation to the component parts of the vehicle. The origin of this frame is located an arbitrary distance above the vehicle's nose. The X and Y axes are each displaced one-hundred inches from the longitudinal axis of the vehicle, and the Z axis is pointed downward, alongside the missile's body. This definition of the MS frame enables the location of all the component parts of the vehicle to be specified as positive numbers with respect to the origin of the MS frame. Center of mass tables, center of pressure tables, and nozzle hinge line distances will all be specified in the MS frame. I had to remember that the MS Z axis is not to be confused with the Ξ axis of the MBF frame.

The Gyro (G) frame specifies the orientation of the stable platform gyroscopes with respect to the Platform (P) frame. Its axes are rotated thirty degrees from the P frame axes. The G frame must be used by MFS to compute the acceleration experienced by the gyroscopes to calculate the platform drift. The higher the acceleration, the higher the drift. This drift leads to a loss of accuracy.

Finally, the PIGA (PIGA) frame specifies the orientation of the three pendulous integrating gyro accelerometers with respect to the stable platform (P) frame. The mysterious PIGA frame is a "scrunched" frame with non-perpendicular axes clustered around the nominal direction of thrust. This arrangement increases accuracy by having all three PIGAs measuring at least a component of acceleration in the most crucial thrust direction. MFS must use this frame to compute the acceleration experienced by each of the PIGAs, and thence for determining the input to the V-loop of the D37D computer. Using the V-loop data, the flight program can correctly guide the vehicle along its reference trajectory. If the IMU is the heart of the missile, then the PIGAs are the heart's life blood.

A force or torque calculated in one frame usually needs to be "transformed" to a different frame. For example, the force produced by a motor needs to be transformed to affect what happens to measurements of velocity by the stable platform PIGAs and attitude from the gimbal angles, as calculated by MFS.

These transformations had to be done following the rules of matrix algebra. In MFS, a matrix is a specially contrived three-column by three-row block of nine numbers. These matrices need to be multiplied together to transform from one frame to another. During the simulated vehicle flight, many of the numbers in these matrices are rapidly changing as integration is being carried out using the Runga-Kutta method.

* * *

It was September 1967. "Ralph, I'd like you to meet Greg Hopwood." Keith White, who worked on control system development for the F-111 fighter-bomber, introduced me to one of his people. "Greg graduated with a math degree from Irvine. I think he might be able to help you out." I knew I was not making sufficient progress. Apparently, management knew that as well.

U. C. Irvine was a brand-new campus located in the middle of what had been the 185-square-mile Irvine Ranch. "What kind of work have you been doing on the F-111?" I asked.

"As you probably know, Autonetics has been flying an airplane low over the islands to determine how well the radar system can see and track the terrain," said Greg in a clear, contemplative voice. "For the past three months I've been programming different parts of the system based on equations describing how they work."

"Well, okay, welcome to the Minuteman Flight Simulator." He was tall and slim, twenty-two years of age. I certainly needed help with some of the math, and he struck me as a pretty smart guy. I set Greg working on programming the matrix transformations, while I continued to focus on the physics part of the problem. We immediately hit it off, developing a deep friendship, spending hours together at the Data Center trying to get our programs to run.

22 A Train Ride to Hell

"FIRE!, FIRE!" I awoke suddenly from a deep sleep, my heart racing. In the middle of the night, acrid smoke had filled the dark cabin. I jolted forward in my seat as a tremendous screeching brought the train to a halt.

"Open the windows!" a man yelled. People in the car were coughing and crying. I stood up, grasped the locks on each side of my window and brought it down a few inches until a cold blast of clear air hit my face. Others did the same, and soon the situation started to calm.

Fifteen minutes later, three men huddled in the snow outside my window, checking the wheels under my seat. One man held a lantern high, another had a flashlight. Light snow was falling; a strong wind flipping the edges of their parkas. One of the massive journal bearings in the truck directly beneath me had lost its oil and had caught fire.

For an hour, we sat there in the cold. Someone said they were waiting for the bearing to weld itself together and seize up, to prevent the wheel from turning. The men re-entered the train, and soon we stuttered ahead, dragging along the frozen wheel. Heavy blankets were passed out. The fire had damaged the heating equipment under the car.

I was wide awake; we were somewhere in Arizona or New Mexico. The white landscape outside shown brilliantly in the light of a waning moon. It was Christmas time, and I had decided to take a week off to go back home to Wichita to visit my parents and siblings. I had boarded the Super Chief in Pomona, where it was running twelve hours late. An epic snowstorm was blanketing much of the Southwest. Little did I know I had headed into the worst snowstorm in Arizona history. Airplanes and helicopters were dropping medicine and supplies for the far-flung native populations. Flying Boxcars were dropping hay bales to the starving cattle.

My blanket was warm, but sleep did not return. Thoughts kept returning to MFS, and how far behind I was falling. During the long night of darkness, my gaze remained fixed on the wonders outside, with occasional worry about the wheel sliding on the rail beneath me.

The train slowed as we passed a herd of deer in the snow, several hundred of them huddled up against the tracks in the moonlight. Later that night, the train crawled and jolted through a series of switches as we slowly passed the burning wreckage of a freight train, cars flipped around every which way. These are sights I will never forget.

It was mid-afternoon when we pulled into Newton, twenty miles north of Wichita, twenty-four hours late. My brother Dave was there to pick me up, having heard about the delay on the news.

It was Christmas Eve, and I was not feeling well. I had caught a bad cold or the flu aboard the train. My illness worsened on Christmas Day, and I had a fever. My sister-in-law gave me some tablets her mother, a retired nurse, had recommended. A few hours later, my fever worsened, and my face and chest began to break out in swollen, itchy, red patches.

Mom called Dr. Lindsley, a kind physician in his early sixties who lived three houses north of us. He came to our house later that afternoon.

"Did you take any medicine?" he asked.

"I took two tablets this morning," I said.

"What were they?"

"I'm not sure. We need to call my sister-in-law." Mom called, and a few minutes later, we had the answer.

"That contains penicillin. You're suffering from a reaction. Have you taken any penicillin before?"

"I probably did when I was a kid," I said.

"Your body is sensitized to the penicillin. Try to rest, and drink plenty of fluids."

"But I have to go back to California on the train tomorrow."

"Son, you're not going anywhere soon. I'll check on you in a few days."

The next day, the agonizing hives covered my face, arms, and upper body. My fever spiked to 108 degrees. I lay in the bed in agony, delirious with pain. "Help! Help me!" The plaster wall beside the bed started to lean toward me and to my horror I could clearly see it was made of brick masonry starting to crack.

"The wall is going to fall on me! I'm going to be crushed! Mom! Mom, get me out of here! Pull me away from the wall!" I didn't realize it, but I was gripped by anaphylaxis as my body's immune system fought against me. I felt weak, delusional, and unable to move—some of the first symptoms of anaphylactic shock.

My parents hovered over me all night like angels, and by morning the fever had gone down to 104. Dr. Lindsley was called, but he was busy with other patients and couldn't come. Over the next few days and nights, in and out of sleep, I began to feel better but the hives had gotten worse, spreading all the way down to my knees.

"It's someone named Bill Herr, calling from Autonetics." My father had answered the phone. "He wants to talk to you."

Mr. Herr was my very senior manager, above Dick Snyder. It took me a long time to get out of bed and make it to the phone in the dining room. "Hello."

"Why aren't you here at work?" asked Mr. Herr.

"I'm sick. I'm suffering from some kind of penicillin reaction."

"Well, get out here just as soon as you can. Everyone is putting out a maximum effort for FTM 201." FTM 201 was the first Minuteman III flight test missile to be launched from Cape Kennedy. All I could do was lie in bed, worry about MFS and hope Greg Hopwood was at least making some progress.

A week went by, and I was feeling better. Dr. Lindsley came by for me. I was lying on the living room couch when he asked me to turn over, giving me an injection of cortisone in my buttocks. The cortisone did the trick. By the following week I was functioning reasonably well. The red hives began to subside, turning a deep purple.

More calls came from Autonetics. This time it was Dick Snyder, a threatening tone in his voice. I worried about MFS, but now I worried I would be fired. *Thank you very much!* Besides having purple skin, my condition was improving. I had survived the brick wall episode, but when I woke up one morning, the bright red hives had finally made it from my knees down to my feet. I didn't know how I could take the train back because it was too painful to even wear shoes.

Nevertheless, I boarded the 5:05 am train in Newton, arriving at Pomona around 7:40 am the following morning. The return trip was uneventful. I slept most of the time. My Chevy was there where I had left it, so I drove to Anaheim, changed my shirt, put on a tie and went to work.

After my three-and-a-half-week absence, Mr. Herr saw that I was back, and called me into his office for a good chewing out. He thought I had lied about my sickness. I simply stood there with my head down and took it—fearful to say much. I had a sudden impulse to rip open my shirt to show him my purple chest, but I thought it would be disrespectful, so I didn't. I wish I had. I left his office with one of my first bad feelings about the place.

23 MFS: In the Thick of It

G REG HAD MADE a lot of progress in my absence. He was a good programmer—very good. The relief I felt at having a partner was amazing.

In my stack of reference documents was a program called CLAMPFS, which stood for Closed Loop Advanced Minuteman Powered Flight Simulator. This early simulator for the original Minuteman was written in 1964. It contained some useful information, but I found the programming style, with its lack of modularity, hard to follow. In any event, the die had already been cast to write MFS for the IBM System 360, more or less from scratch.

From my earlier experience, I came to mistrust what programmers sometimes refer to as "spaghetti" code, long hard-to-follow programs containing multiple unrelated program routines. Instead, wherever possible, I decided to break the code into smaller pieces that are easier to understand. This practice, however, introduced some performance penalties because of the overhead required when one routine had to call another.

Thousands of numbers described the missile and its physical environment. The entire group of numbers was classified SECRET. Many of these numbers were constants, for example, mass and location of various components, expressed in the Missile Station (MS) frame. Tables of constants described such things as wind velocity versus altitude or thrust versus time for a given stage.

Other numbers were expressed as matrices, including coordinate frame transformations. Many of these numbers had been obtained from careful experimental investigations performed at the various companies that made the missile components. Most important were the numbers specifying the reference trajectory for a given mission.

All the numbers had associated names. It was a daunting effort to link these names with the different names used in MFS referring to the same values. This was eventually done through what amounted to a huge table I started calling the "TRW-MFS Rosetta Stone." The name was an homage to the original 196 BC stone inscribed with a decree in three different languages. The source for many of these numbers was a company called Thompson Ramo Wooldridge, Inc., or TRW for short. This part of TRW, Space Technology Laboratories, (STL), in Redondo Beach, California, was instrumental in supporting the Air Force's ICBM work, including Minuteman III. Other numbers were coming from discussions with Autonetics engineers.

* * *

Minuteman Flight Simulator can be considered a simulated computer interacting with a simulated environment. It simulated the program of the D37D flight computer within its IBM System 360 host computer. Many times, during the missile's flight, the onboard computer commands events to occur which actually will not happen until some later time because of system delays. The simulation must somehow remember that an event is pending, and then return control back to the simulated computer (DSIM) so it can continue its instruction-by-instruction operation.

When the delay time elapses, DSIM had to be interrupted to allow MFS to simulate the event. This could even happen right in the middle of a simulated instruction. For example, when a protective skirt is commanded to be jettisoned from one of the rocket stages, some kind of fuse is lit. After lighting the fuse, the D37D goes on making its calculations blissfully unaware of what happens next. A short time later, tension members holding the skirt explosively detonate causing the skirt to fall away. The instant this occurs, the vehicle becomes lighter, and MFS must model the sudden change in mass and aerodynamics.

Any sudden change is termed a "discontinuity" in the differential equations being solved by Runge-Kutta. To handle these events, MFS has to integrate up to the exact time the discontinuity occurs, print out the status of everything, simulate the event that caused the discontinuity, then print out everything on the other side of the discontinuity and resume integrating.

I walked over to the adjacent bullpen. "Greg, I am having a hard time dealing with how to remember and process future events," I said. "In particular, I'm worried about handling a situation where I have to process a pending event, when another event becomes pending before the first event is serviced. I don't know how to handle this kind of thing. Any ideas?"

After thinking about the problem for an hour or so, Greg came to see me. "Maybe we can use a push-down stack," he said.

"A what?"

"You know, a push-down stack. Have you ever been in a cafeteria?"

"Well, yes, of course."

"Sometimes there is a spring-loaded device holding a heavy stack of plates recessed in the countertop. When you take a plate off the top, the reduced weight lets the stack move up for the next person in line. The last plate going in the stack is the first one out. Batches of plates are added to the stack by the cafeteria workers. Plates are taken out by the customers. Anyway, stacks are kind of a new concept in computing. I think we might be able to devise some kind of push-down stack to process events in MFS."

"I never liked eating in cafeterias," I said.

It took Greg a few days to implement his idea that he called a "Time Interrupt Push-Down Stack." Instead of dinner plates, there were "events." Each event had three elements. The first was the interrupt time, expressed in computer word times since the computer began operation, with each word time taking exactly 78.125 microseconds, the basic pulse of the D37D.

The next element was a so-called *flag* designating the physical action to be simulated, for example, a skirt jettison. The third element was the contents of the D37D D-loop at the time the event is pushed onto the stack. Bits in the D-loop specify which of the PBPS thrusters should be turned on or off.

Greg explained how this would work. "Along with all the events being pushed onto the stack, there is a thing called a *stack pointer* which gets reassigned to point to whichever event will occur first. When the interrupt time is reached, the event data is 'popped' out of the stack to make room for subsequent events. Many different MFS program routines can push events onto the stack, but only the master timing routine can pop events back out."

* * *

Greg and I were hard at work developing program routines that would handle many important details concerning the so-called "boost" phase of the flight, that is, the first three minutes or so, during the operation of the three solid propellant stages. Voltage outputs from the D37D H-loop control the first stage nozzle angles and the positions of precision motor-driven pintle valves for the second- and third-stage liquid injectant thrust vector control.

As soon as the nozzles or pintles "see" new voltages, they begin to move. But because they cannot respond instantaneously, they take a finite time to reach their final commanded positions. Each has a detailed response function that needed to be simulated. Moreover, the injectant, besides providing steering, additionally augments the propulsive thrusts from each motor, so we needed to take this into account as well. A "shims" model took into account imbalances in the vehicle mass distribution causing the motor thrust forces to not act exactly along the centerlines of the stages.

We created routines implementing the effects of the atmosphere on the boost vehicle up to an altitude of three hundred thousand feet, including lift and drag forces. We implemented the Sissenwine Wind Profile simulating wind shear conditions in the lower atmosphere. Wind shear can cause aerodynamic torques that could throw the missile off course. Most of these calculations involved table look-ups of data supplied from wind tunnel tests. As the vehicle ascends, gravitational attraction decreases which had to be simulated as well. We used a gravity model involving the first and second so-called "zonal harmonics" for the Earth. These data had been derived from years of careful observations of satellite orbits.

During the boost phase flight of the vehicle, the D37D issues many digital on/off commands called "discretes." Some of these discretes turn on gas generators, batteries, and other devices. For these discretes, no action needs to be taken by MFS; only a message should be printed on the output listing. Others, such as motor stage ignition, roll motor control, nose shroud removal, and inter-stage skirt jettison require Greg's Time Interrupt Push-Down Stack, and the simulation of abrupt changes in vehicle mass and aerodynamic properties.

Thrust termination for the first two stages is based on elapsed time and acceleration. For the third stage, it is based on the vehicle velocity. A short time prior to termination, the D37D enters a *fine countdown mode*. To do this, the flight program loads the U-loop and the C-loop (another high-speed loop) with specific numbers related to all three components of the velocity to be gained before termination.

The value of the U-loop needed to be updated by MFS using a relation involving the C-loop values and PIGA velocity increments. When the flight program senses the U-loop transitioning to a negative number, it issues the third-stage termination discrete. After a delay, six radial ports at the top of the motor are blown out with explosives to cut off its thrust and the tension members fastening it to the fourth stage are explosively severed. This inevitably causes a bang and transient force that also needs to be simulated.

* * *

As we received more data from TRW, our attentions began to focus on the Post-Boost Vehicle (PBV), comprising the Post-Boost Propulsion System (PBPS), the Post-Boost Control System (PBCS), and the warhead mounting platform at the nose with its decoys, chaff, and the three Mark 12 RVs. In some respects, simulating the physics of the PBV was a bit easier than the three-stage boost vehicle, since it was assumed that the PBV would operate above the upper limits of the atmosphere where aerodynamic effects could be ignored.

We needed to simulate the actions of the PBPS axial engine as well as the ten small attitude control thrusters, any group of which could be operating simultaneously. The design showed four pitch thrusters, four roll thrusters, and two yaw thrusters located strategically around the outer surface of the PBPS. The location and orientation of the axial engine and each of the ten thrusters were expressed in the Missile Station (MS) frame but the force magnitudes and directions had to be continually transformed to the Missile Body Fixed (MBF) coordinate frame by using matrix multiplications. The total force acting on the PBV is the sum of the forces from the axial engine and the forces from the attitude control thrusters. Additionally, there are reaction forces caused by jettisoning a re-entry vehicle or decoy, reaction forces from dispensing chaff, and the sudden shock force from the third stage thrust termination event.

The fuel and oxidizer are stored in flexible bellows within separate tanks. As propellant is used up, the bellows collapse under gas pressure from a helium bottle, and the bellows' spring force increases. This causes a steady decline in propellant pressure during the flight, with a resulting decline in thrust output. We simulated this "bellows effect" with a decreasing straight-line function of the remaining propellant.

The nominal 315 lb.-thrust axial engine can move up to plus or minus thirteen-degree angles in pitch and yaw directions, and must move through a large angle particularly after deploying an eight-hundred pound RV to maintain its line of thrust through the PBV's center of mass. The angles are commanded by voltages derived from the D37D H-loop. Since the engine cannot change direction instantaneously, it was necessary to simulate its response to the commanded voltages. Further, because of the manner in which the engine is gimballed, a pure yaw command will introduce a small amount of pitch, and vice versa for a pure pitch command. It was necessary to simulate these effects as well.

The thrust force from the axial engine is simply turned on or off by a discrete output from the D37D. When thrust is turned on, there is a short delay before thrust begins to build up to its maximum as determined by the bellows effect. Similarly, when thrust is turned off, there is a delay before thrust decays to a level near zero, followed by a longer decay to zero as residual propellant in the engine burns away. Simulating this behavior made good use of the Time Interrupt Push-Down Stack.

Simulation of thrust forces from the ten small attitude thrusters is again, simply an on/off affair plus time delays. The thrusters are turned on or off depending on bit settings in the D37D D-loop. There is, however, another factor that had to be taken into account. When the axial engine is thrusting, it causes a pressure drop in the propellant lines feeding the attitude thrusters, reducing their thrust by a few percent on top of the reduction due to the bellows effect. I thought this was like being scalded in the shower when someone in the next room turns on a cold-water faucet.

The sum of all the thruster and axial engine forces as well as any object deployment forces feed into the PBV's Translational Motion Model. Similarly, the sum of all torques produced by the thrusters and axial engine and deployment torques feed into the PBV's Rotational Motion Model. Torques were computed for each thruster and the axial engine by multiplying their forces by their distances from the PBV's center of mass (CoM).

The Post-Boost Vehicle must maneuver in space over a range of attitudes and thrust from its axial engine to deploy objects destined for up to three different targets. It was necessary for MFS to keep track of these deployments and the reduction of mass associated with them. The structure of the PBV and each deployable object on the PBV has a set of ten mass properties associated with it. There is the mass of the object, the X, Y, and Z components of the object's CoM expressed in the Missile Station (MS) coordinate frame, the X, Y, and Z "moments of inertia" of the object, and the three "products of inertia" of the object. Each stationary item and each deployable object on the PBV is characterized by its own set of ten values.

I knew that the moment of inertia of a rigid body is a quantity that determines the torque needed for a desired angular acceleration about a rotational axis. (This is like how the quantity "mass" determines the force needed for a desired acceleration.) It depends on the body's mass distribution and which axis about its center of mass is chosen.

The "products of inertia," a more difficult concept to understand, come into play when the object lacks symmetry in one or more directions. The moments and products of inertia of an object are most conveniently expressed as six values in a three-by-three matrix called an "inertia tensor." Given the collection of stationary items and deployable objects on the PBV, each with ten numbers, a mathematical theorem called the "Parallel Axis Theorem" was used to mash all these numbers together to produce a single set of ten numbers characterizing the combined mass properties of the PBV.

Before the simulated launch, the PBV Structure Model will be loaded with the mass properties of all its parts from the TRW-supplied data, including the consumable items like fuel, oxidizer, and chaff. It was up to Greg and me to create program routines that would compute the combined mass properties of the PBV and to continually recalculate them as the fuel, oxidizer, and chaff are consumed, and also every time an object is deployed. Weeks passed before we had generated several pages of flow charts describing the logic and physics of all this.

Decoys (smaller dummy RVs) are mounted on platforms which can be rotated downward in the pitch plane of the reference trajectory. The mass property numbers supplied by TRW are for the rotated condition, so before the simulated launch, we had to "rotate" their inertia tensors back to their pre-deployment launch attitude state. Another of the dozens of details we had to take into account.

I was trying to get my head around yet another issue. The three-by-three inertia tensor of the entire PBV can be thought of as an ellipsoid centered at the PBV CoM. This ellipsoid has three principal axes at right angles to one another: a major axis along the direction of the greatest inertia, a minor axis in the direction of the least inertia, and an intermediate axis with a middle-of-the road inertia. I knew from physics, that motion around this intermediate axis is unstable. *What will happen when the PBV pitches down around this axis to deploy an RV?*

* * *

I needed to go home and get away from Autonetics. It was a pleasant evening as I drove west on the Riverside freeway thinking about the unstable axis problem. If you flip a tennis racket a full rotation about its unstable axis, it will execute a twist in the air and land back in your hand with the original upward-facing side now facing down. It does this flip in the complete absence of any external forces or torques. *How can this happen?*

Suddenly, out of the clear sunset-lit sky, I had the answer: Of course, you idiot, you have to use Euler's equation of motion to calculate motion about the unstable axis.

Overjoyed by my epiphany, I was vaguely trying to recall the math when the Chevy sputtered and ran out of gas. I pulled off to the side and began to laugh. That was the solution for MFS! If the PBV rotation includes motion around its unstable axis, it will twist in space like the tennis racket. Euler's equation will properly simulate this action. With this insight in mind, I began cheerfully walking alongside the freeway toward the nearest exit with the hope of finding a filling station.

* * *

Another week of programming and testing had finished the proper Euler calculations. We still had a long way to go. As the PBV rotates to different angles in space for object deployment, the effect of these rotations on the IMU Model's

187

three-gimbal stable platform needed to be simulated. Two main issues concerned us. The first was exceeding a gimbal's range of motion, slamming it into a stop. The second was so-called "gimbal lock," where the inner and outer gimbals become aligned with each other causing the Platform (P) frame to become undefined. Either of these conditions could happen if the flight program mistakenly does a foolish maneuver. Knowing the Flight Weenies, I was sure this could occur, so, we needed to check for these conditions.

The Minuteman Flight Simulator was in pieces, and the card deck was running at over three-thousand cards—too heavy to carry to the Data Center, so we schlepped them back and forth in the Chevy. In mid 1968, at only a half-dozen runs a day, we were falling behind and desperately needed help. In truth, eighty hours a week wasn't enough, and I was starting to go crazy.

* * *

Dick Capello, who oversaw our work, must have known we were falling behind. Summer had come and the first launch from Cape Kennedy was coming up in a few months. Greg and I were churning out code, and thankfully large chunks of MFS were showing signs of working. But there was no doubt we still needed help.

"Ralph, I'd like you to meet Larry Hambly," said Dick. "He has a fresh degree in physics from Montana State."

"Okay. Hi Larry. It will be good to have a fellow physics guy on the project." Larry looked young and eager to get started on something. I soon found he had a pleasant personality and even disposition.

We had been using a simplified zonal harmonics gravity model for MFS, the so-called "J and D" model from work done in 1960. In truth, the Earth's gravity, at a small scale, exhibits quite a bit of lumpiness. If there is a mountain range near the launch site, the mass of those mountains will pull a little on the missile. We needed to take these effects into account. A much more precise model, based on so-called "eighth-order spherical harmonics" was needed. TRW had recently supplied the numerical values for such a model, which had been derived from the most recent careful measurements of satellite orbits.

I put Larry in charge of programming the spherical harmonics model, dumping a stack of papers with the relevant equations on a vacant desk behind me. "Let me know if you have any questions," I said. Looking a bit bewildered, he started to peruse the material. The basic idea was to calculate the Earth's potential field in the Earth-Centered Earth Fixed (ECEF) frame. This would involve computing a number of trigonometric functions and so-called "Legendre polynomials" in spherical coordinates to find deviations in

188

gravitational potential from that of a simple spheroid. The model used so-called "recursion relations," to save precious computer time, for computing higher-order trigonometric terms up to the eighth order.

Once the gravitational potential is computed, its so-called "gradient" needed to be computed to find the gravitational acceleration at a point in space. Finally, the gravity value needed to be transformed from the ECEF frame to the Launch-Centered Earth Fixed (LCEF) frame. (The position of the missile at any point in time is computed with respect to this latter coordinate frame.) Relieved to hand this off, I refrained from mentioning to Larry that I had received a D in high school trig.

Technically, Larry couldn't start working on the new gravity routine until his security clearance came in. I urged him to start anyway, since time was so crucial. He would arrive at 8:00 in the morning, work steadily, and hang it up at 5:00. Before leaving, he would clean off his desk, and line up his pencils in a neat row.

* * *

I put Bob Moffitt to work learning FORTRAN programming. I thought once he learned how to program he could help me out with writing some routines for MFS.

Betty Jo Kraus joined our small team. She was a bit older, and wiser than I, and was a pretty good programmer. She had a cheerful smile and pleasant disposition. I thought we needed a way to go beyond the hundreds of pages of printed output from MFS. We needed some form of graphical output as well to make it easier to spot subtle instabilities in motion, particularly for the post-boost portion of the flight and object deployments.

The project in at-sea gravity measurements had introduced me to the Stromberg-Carlson 4020 cathode ray tube printer. I put Betty Jo on the task of writing a post-processor program for reading the MFS output and turning it into graphs using the 4020. She soon made some progress, dubbing her contribution "VIDEO," for Validation and Demonstration. Modifications were made in MFS to create a magnetic tape containing all the nearly seven hundred types of variables and messages output from an MFS run. This tape, as well as user-specified commands, formed the input to VIDEO.

To handle the user commands, she began to devise a special VIDEO input language allowing users, like the flight programmers, to select which variables were to be graphed, at what time during the simulated flight they were to be graphed, and the type of graphing to be done. Special on/off plots of the PBPS axial engine and attitude thrusters needed to be devised. This was all turning out to be a monumental effort, since the SC4020 had few built-in program routines.

* * *

With Betty Jo and Larry off and running, Greg and I began to tackle the complexities of the warhead end of the beast. We needed to simulate the re-entry system in considerable detail, including its logic, time delays, and the deployment reaction forces and torques on the PBV.

The re-entry system contains an electromechanical stepper switch with three poles and nine positions. A discrete output wire is connected to each of the three wiper poles on the switch. The output wires are connected to different parts of the re-entry system at each position of the switch. This piece of hardware is a bit like the old-fashioned telephone relays back in the days before electronics. The switch starts at position one at the beginning of flight and is stepped forward one position each time a stepper-switch reassign discrete is issued by the D37D.

There is also a decoy stepper switch having a single pole and twelve positions. A D37D discrete output wire is connected to the switch wiper and each stepper position has an output wire connected to a decoy release mechanism. This arrangement serves to jettison the decoys, one by one, throughout the flight. The switch starts out at position one and is stepped forward by another discrete output from the D37D.

Decoys are mounted on helical-grooved posts on rotatable platforms on each side of the PBV. The decoys are cone-shaped bodies mimicking the aerodynamic properties of the RVs. Prior to jettisoning, the platforms rotate from their launch attitudes to their jettison attitudes. This changes the decoys' centers of mass with respect to the PBV's CoM and also changes their inertia tensors. We needed to simulate these effects as well.

To determine when and where a decoy will impact on the Earth's surface, its deployment condition (time, velocity, and position) must be precisely known. As a spring-loaded decoy leaves its post, it rotates and translates, imparting a reaction torque and force on the PBV. We needed to simulate these effects. The decoy's spin keeps its attitude stable while in free-fall and during its re-entry in the atmosphere.

The re-entry system also has a pair of chaff dispensers, one on each side of the PBV. Chaff, numerous small wires of various lengths designed to reflect enemy radar signals, is stored in each dispenser in lumps referred to as "boats."

Not knowing exactly how these dispensers worked, I visualized the boats sandwiched between rolls of thin plastic film, something like Saran Wrap. I assumed a pair of rollers acted to roll up each of the films to release the chaff into space. Regardless of how it worked, the chaff could be dispensed at a pre-determined flow rate and speed. Like for the decoys, we needed to simulate its reaction forces and torques on the PBV.

The chaff boat feed rate and velocity are set by D37D voltage outputs and the position of the re-entry system stepper switch. Two separate time delays needed to be simulated, associated with the time a boat hits the roller mechanism and also when it leaves the roller mechanism. These delay times are a function of the chaff ejection speed. Like the decoys, chaff puffs needed to be followed by MFS to the top of the atmosphere, requiring knowledge of precise release conditions. After a chaff puff departs the PBV, the flight program must exercise caution to avoid disturbing the puff with the exhaust plume of the PBV axial engine.

The final consideration we needed to think about was releasing the three rounded cone-shaped Mark 12 RVs, the business end of the beast. To release an RV, the flight program must first pitch the PBV downward in the pitch plane of the trajectory to a near straight-down attitude. While in this attitude, a special flight program Deployment Attitude Guidance function zeros out the velocity to be gained before release, using the small PBV attitude thrusters. This maneuver orients the RV in space so that when it re-enters the atmosphere above the target many minutes later it will have the correct zero-lift attitude with respect to the air mass, nominally sixty-three degrees from the vertical.

Each of the three RVs, arranged in a triangular configuration on the payload mounting platform atop the PBV, is held tightly in place by three zero-impulse ball locks. Besides these mechanical connections, there is an electrical connector interface with the D37D. If the flight program decides that all is well when the velocity to be gained crosses zero in all three axes and there is no rotation, a code is sent to the RV, starting its warhead arming procedure. Otherwise, the warhead will not be armed. Then, a discrete output is issued through the re-entry system stepper switch to release the ball locks on the RV.

Since the ball locks do not cause any reaction forces or torques on either the RV or the PBV, there is nothing for MFS to simulate. The flight program uses the yaw thrusters to back away from the released RV, taking care not to bump into a neighboring RV a short distance away on the mounting platform. Seconds later, the RV activates its hot gas generator spinning it up to several revolutions per second and it is on its way.

Using its axial engine, the PBV can rocket over to two more places in "velocity space" to release a second and third RV on precise trajectories to their targets. The spread of the destructive footprint is determined by whether the deployments happen earlier or later in the flight. It is all over in less than nine minutes from stage one ignition. Importantly, MFS needed to faithfully execute all the maneuvers commanded by the flight program, whether they made sense or not.

After MFS simulates the deployment of decoys and chaff, and releases the RVs, it is time to compute their ballistic trajectories. In another ten or fifteen minutes the chaff puffs will hit the top of the atmosphere and the RVs and decoys will hit their targets. This part of the simulation is quite simple, since gravity is the only force acting on the objects and it is sufficient to treat the objects as point masses.

The General Electric Mark 12 RV has forward, aft, and mid-sections joined together by breech lock threads. The mid section houses a W62 thermonuclear warhead with the explosive power of 170 kilotons of TNT. Lawrence Livermore National Laboratory had begun the design of the W62 in 1964. The main external surfaces of the RV are made of ablative carbon-phenolic heat shield material, and the nose tip is protected with a Teflon boot.

The shield is designed to achieve minimum radar cross section and to protect the interior from the blazing heat of re-entry. The hot gas spin system, located in the aft section, stabilizes the reentry vehicle in its correct re-entry orientation after release from the PBV. The RV contains the arming and fusing assembly providing various height-of-detonation options.

I did not know it at the time, but the development of the Mark 12 had been fraught with difficulties, both politically and physically. There had been opposition to creating a weapon with such a small yield, even if there were three of them, versus having a single high-yield warhead.

The Mark 12 has a high "ballistic coefficient" equal to its weight divided by the product of its effective cross-sectional area and drag coefficient, the latter term being a function of the RV's shape. An RV with a high ballistic coefficient has reduced susceptibility to wind and weather conditions present at the target. Early test reports revealed excessive erosion of the shield material during re-entry, causing asymmetric ablative patterns to form that led to wobbling motion or even spin reversal, resulting in a loss of targeting accuracy. Fortunately, we did not have to worry about these details for MFS.

The final calculation to be performed by MFS must compare the simulated impact locations with those of the reference trajectories, results of great interest to the flight programmers. Of course, a simulation is only a simulation. For real missiles, there is the notion of Circular Error Probable (CEP) which is a circle centered at the target wherein fifty percent of the RVs strike. In many cases, the pattern may be an ellipse, elongated in the direction of the trajectory, indicating under- or over-shoots caused by errors in the release velocity. Cross-range errors are often due to platform alignment errors at launch.

A new fellow, Dave Dorius, started helping. I put him in charge of programming many of the "details," including the freefall programming — following the deployed objects all the way to the targets. Dave had been a captain in the Air Force and had gone to flight school. That hadn't worked out so he switched to data processing, having recently joined Autonetics. He was perfectly dressed and quite formal; someone I never actually got to know.

It occurred to me that I was now in charge of a group of four other people, all working to make the Minuteman Flight Simulator a reality. Besides my immediate crew, there was Bob Best in charge of DSIM, Al Sheue, and Bob Moffitt, who had written a few small routines and was helping out with taking MFS runs to the Data Center. It had a familiar feel, not unlike the Propulsion Research group I had founded back in high school. It took effort to manage personalities and schedules as well as the mathematics and physics problem at hand. Except for my brother Dave, I had lost contact with all the others that had worked so hard to build and launch our rockets.

We were still far away from assessing whether a flight program could guide the missile to anywhere near hitting the targets. "Dave, are you making progress on the freefall programming?" I asked. "I noticed you weren't here most of the afternoon."

"I put in my time the previous night. I guess you don't realize I'm taking classes at Fullerton to get a master's degree in mathematics."

"Okay, I just want you to realize that your routines are a vital part of MFS. Hopefully you're coordinating with Larry on the new gravity model."

* * *

Greg and I got into the habit of driving the MFS program boxes to the Data Center, bribing an operator with coffee to get priority and then hanging around for an hour to check if the decks had been successfully read into the computer. Since the runs were taking more than an hour to complete, we had time to drive a few miles west to Carrow's twenty-four-hour diner. Many times, we ate their satisfying hamburgers and French fries at midnight or 1:00 am before returning to pick up our cards and printouts. It was a good time to get to know one another.

One time, the operator told us he had several high priority jobs ahead of us. It would be another two hours before he could run MFS. The decks were still in my car, and we were pretty hungry, so we decided to head for food. A mile from Autonetics, the Chevy sputtered and quit.

"Oh crap," I said. "I can never remember to get gas."

"What should we do?" asked Greg.

"Lock up and start walking."

We left the three boxes of the MFS program, complete with all the secret Minuteman III data, flight program deck, and TRW reference trajectory in the back seat of the car. We reasoned that even if someone found the car alongside La Palma Avenue, they wouldn't figure out what was inside, and even if they saw the card decks they couldn't grasp their significance.

After walking together for thirty minutes, we lucked into an all-night gas station, bought a gallon of gas, and headed back to the car.

"Don't walk so fast," Greg said. I slowed when I remembered he had something wrong with his knee.

All seemed to be untouched. After returning to the station and gassing up, we parked at Carrow's outside a window where we could keep an eye on the car. After ordering our usual, I said, "What do you think about all this?"

"Not much," said Greg, after a pause, "What do you mean?"

"I mean, we're doing important work. It's also pretty interesting," I said.

"Yes, I'll give you that," he said softly, staring out of the window next to our booth. The car was there, with the decks safely intact.

"Too bad we can't tell anybody about our work. I can't wait to finally tell my father what I've been doing. My brother as well."

"Yeah," he said.

I asked him about his family living in Tustin. He told me his parents met in the Navy during World War II. "My dad is working on the MOL project at Douglas Aircraft. MOL is the Manned Orbiting Laboratory. It's a secret Air Force project." Greg told me about his sister Vicki, and brothers Chris and Mike. Chris was sixteen and was training a falcon. I told him about my younger brother David, sister Karen, and baby brother Brian.

"When I was a year old we moved to Santa Fe, New Mexico," I said. "Around age three we moved to Wichita. My father is an architect. When I was very young growing up in Wichita, he worked for Boeing." We finished our burgers and fries, drove back to the Data Center, and submitted the job. It bombed five minutes into the run. We called it a night.

* * *

John Hlavac and some of the other Flight Weenies, my customers for MFS, were starting to request all manner of user input capabilities going way beyond the physics of the problem. They wanted to be able to introduce different kinds of perturbations like extra PIGA counts and gimbal angle counts, checks for gimbal angle limits, and others. They were being pressured by their boss, John Sweet. The formidable Mr. Sweet, who worked for Dick Snyder, had a personality I would not at all consider "sweet."

They requested a number of new features, including a checkpoint restart feature, turning on/off debugging printouts, extracting and printing the contents of specified D37D memory locations, and performing velocity-to-be-gained calculations for comparison with the flight program.

My '56 Chevy was on its last gasps. I traded it in for a new 1967 British Racing Green MGB sports car that was sitting on the lot. Driving around in that zippy little car with my Joe Cool sunglasses and with the top down gave me an intense feeling of pride.

Lacking a TV set in my apartment, I had been largely unaware of the world outside of work. I listened to my MGB's radio going to and from Autonetics. The war in Vietnam had entered a new phase with something the news called the Tet Offensive. President Johnson had suspended the B-52 runs over North Vietnam and had made the stunning announcement that he would not run for reelection. Meanwhile, it seemed that the Brezhnev regime was building a massive inventory of nuclear missiles. Time was quickly running out for the U.S. to mount a retaliatory strike force.

It was around midnight when I got to my apartment, climbed up the stairs, and laid down in my bed. The hours went by staring at the ceiling as a deep worry washed over my body. I was so tired I could not sleep or even close my eyes. I had been concentrating mostly on the physics and logic of the MFS program and was feeling completely overwhelmed. Greg was working full time on logic and sequencing, Larry was still trying to implement the new gravity model, Dave was wrapping his head around the freefall program, and Betty Jo was starting to get somewhere with graphical output. Bob was still learning to program in FORTRAN. How could we possibly fulfill these new requests from the Flight Weenies? No sleep came. It was seven, so I dressed myself, and returned to work.

I began to think of myself as a robot, aimlessly hacking out code for the brain and heart of the new missile. I longed for Frankie's companionship and resolved to strike up a new relationship with someone.

* * *

Georgie Girl was a popular song and movie. My own Georgie girl was a keypunch operator and friend of John Hlavac. Earlier on, she had punched some of my card decks. Blonde, with blue eyes, she was very nice. One day, she invited me to go listen to her friend sing some western music at a local bar. I picked her up in my MGB from her apartment adjacent to Disneyland. After the evening concert, I drove her home, and kissed her on the lips when we parted.

From that night onward, we enjoyed being together, taking long rides, especially along the coast highway to Laguna and San Clemente. Sometimes we watched the nightly fireworks from her balcony. I bought her an ice cream cone at the drugstore; when I pulled a few bills out of my wallet, she glanced at my Kansas driver's license. On it was my status: "married."

"You're married?" she asked, glaring at me.

"Well, not exactly. We're divorced now. Just haven't thought to mention it to you."

"Drive me straight home." When we arrived at her apartment, she slammed the car door shut and ran up the steps. *I am so stupid.*

I thought it was the end of our relationship, but fortunately, we got past it. While driving with the top down in the MGB her long hair would flip about, and I would reach over and give her a fond squeeze above her left knee. I always enjoyed her company, but we never entertained the notion of having sex. I wasn't ready for that, but her companionship eased my loneliness. We would browse the local Heathkit showroom where I would point out this or that piece of test equipment I needed. In my apartment, we sat on the couch and talked. I would tell her about the progress I was making with my kitchen table computer. Georgie was a welcome distraction from work. I needed to finish MFS, but I also needed to live a little.

24 MFS: Flight Test Missile 201

I WAS GETTING MY SALARY with overtime pay up to sixty hours a week but putting in around eighty in my windowless environment. Any work time above sixty was called "green time" for some reason. You were not paid for green time. Recently, we had all moved from building 72 to the much larger new building, 235. The company was hiring rapidly with new strategic programs and technology for the escalating war in Vietnam. Autonetics was busy hiring so many people that no vacant offices were left in which to play the battleship coffee game. And no one in the building had a minute of extra time.

Often, I would get started by 8:00 am, concentrating ferociously on thinking, programming, and punching cards. At around 6:00 pm, the place would get strangely quiet. *Oh! I guess people are going home.* Betty Jo would leave around then, and Greg would call it quits by 10:00 at night or by midnight. The place became devoid of human sounds for a while, then I started to hear voices. *Oh! People are coming in. It must be morning.*

News in the Autonetics Skywriter said that the company was planning a gigantic new building to house the bulging workforce. The new building, shaped like an ancient Mesopotamian temple, was under construction far south of Anaheim, in Laguna Niguel. Jerry Brown—aka Mr. Wizard—had bought a house in nearby Mission Viejo, speculating that housing prices would soon skyrocket. He moved into the house several months before but had mentioned he was having trouble with his neighbors.

Even though I felt guilty being away from work at all, I decided to take a few hours off to visit Jerry and his new house. It was a pleasant Saturday afternoon drive to Mission Viejo. The house, in a large development of nice homes, had three-foot-high weeds growing around it. Jerry answered the doorbell. "Hi Jerry, what's with all these weeds?" I said, gesturing in the general direction of what should have been a lawn.

"I guess I need to buy a lawnmower. Come on in." The bare interior of the house had a pad and sleeping bag on the floor. All the windows had been painted over with white paint. Noticing my dismay, the ever-practical Jerry said, "I guess you are supposed to have curtains. But this does the job." Sunlight filtering through the paint bathed the living room. He offered me a glass of tea.

"Thanks, Jerry." It was a hot day, so we stood in the empty kitchen and each had a cold glass of tea. I was in awe of this brilliant but unconventional man.

* * *

All alone in building 235 at 3:00 in the morning, I was not feeling well. A diarrhea attack was coming on, and I had to run to the corner restroom. I had been having trouble with my stomach, and recently had taken time to see a doctor. He told me I was eating too much meat and needed to start eating more vegetables. My apartment abutted a Ralphs Supermarket full of vegetables. I went there and bought a few carrots, celery, and some bananas, but I didn't have any pots and pans and had no idea how to cook vegetables. I probably stayed for fifteen minutes in the restroom before returning to my desk. To my horror, the two boxes containing several thousand source cards of the Minuteman Flight Simulator were gone!

A feeling of panic enveloped me as I looked around my cubicle. Did I put them on someone's desk? I walked around to the adjacent cubicle where Greg sat. Nothing. Think. Where did I put them? I looked under a half dozen desks. Then, I remembered I had set them on the floor next to my desk to make room for a stack of manuals I was studying. My God, the cleaning crew must have picked them up in their nightly sweep!

I ran up and down between the aisles of the immense bullpen of cubicles. A guy in a uniform! "Did you pick up some boxes of cards back there?" I asked, pointing in the general direction of my cubicle.

"Were they on the floor?" he said.

"Yes, I think they may have been."

"Well, then they're gone now. Likely in the dumpster."

I ran down the stairs to the exit. The dumpster was at the loading dock near the door on the south side of the building. I hoisted myself up on the dock and back down into the enormous bin, halfway full of paper listings, piles of cards, and dozens of boxed card decks. I tried to remember if this was the night of the shredder. I had used the computer to punch some backup decks that were safely back at my desk, but it had been three weeks ago and those cards lacked any text along the top edges.

Balancing on the card boxes, I began to remove their tops one by one in the meager illumination coming from the loading dock. A light rain began to fall. I needed to hurry up. By the time they were found they would be ruined.

The missing MFS boxes were buried only a foot deep in listings and loose cards. All appeared to be in order, judging from the diagonal magic marker stripes across each deck in the boxes denoting the individual program routines. Hefting the boxes up; I pushed them onto the dock and climbed out. I looked around in the darkness through the softly falling rain, but the monster shredder was nowhere in sight.

* * *

Dire news was coming out of U.S. intelligence efforts to understand what was happening in the Soviet Union. Recent photographic evidence showed a massive buildup of ICBM sites beginning all over their country, mostly in hardened silos. At roughly the same time, the Soviets were said to be developing an anti-ballistic missile (ABM) capability that could defeat our fleet of Minuteman I and II missiles. Minuteman III with its MIRV capability was intended to overwhelm the enemy ABMs with a second strike following a Soviet attack on the U.S.

I thought about when I was a student at Wichita's Southeast High School, when the mournful sound of air raid sirens was a continual part of our lives. In those days, we all felt there was a good chance we'd be nuked. A large storm drain emptied into a creek not far from the school. Phil, Steve and I had explored it northward for a quarter mile where we encountered a big concrete room with several smaller pipes leading into it. In this underground room the blast might just scrape off the veneer of civilization and pass over our heads. In our minds' eyes we would play the drama forward, planning on where and how we would get food and water after the blast. Hawkins had pointed out there would be a problem if it rained, and we had laughed. Steve always pointed out the flaws.

Boeing and McConnell AFB were surely targets. Now, it occurred to me that a place like Autonetics would also be a juicy target. I had traded one bull's eye for another.

"This turkey is on Clark Clifford's blackboard," said a worried-looking John Sweet. He had called a meeting of the half dozen flight programmers along with the five of us working on the Minuteman Flight Simulator. President Johnson had recently appointed Clark Clifford as the Secretary of Defense, replacing Robert McNamara. "TRW has scheduled the launch of FTM 201 in August, earlier than we had thought. Everyone has to put in a balls-out effort."

Mr. Sweet went on to talk about the inclusion of several critical flight program routines, emphasizing the need to conserve memory in the D37D. Hlavac rolled his eyes, glancing at me as if I was Bernstein, starting to do their silly Mutt and Jeff shtick, eliciting an almost-chuckle from me, followed by a quick glance from Mr. Sweet. "I'm deadly serious about this," he cautioned. "When I say balls-out effort, you know I mean balls-out effort," he said, looking straight at me.

What exactly is a balls-out effort? I wanted to ask, but I was afraid to. Of course, MFS was the only way to determine if the flight program actually worked correctly, short of the expensive launch of a real missile. But what if the design or construction of MFS was flawed? I left the meeting with an even sourer feeling in my stomach. I knew MFS still had a long way to go. I didn't think I could work any harder than I already did.

* * *

"Ralphie, are you hearing that?" asked Hlavac. The two of us were driving to the lunch place on Imperial Boulevard. I turned up the volume on the radio. Something called *MacArthur Park*. It was a new song climbing the charts.

"Who's singing that?" I asked.

"I heard it was an obscure actor named Richard Harris. I predict this song will be number one!" said John, gleefully. For John, I think the song might have subconsciously appealed to him as a metaphor for coding the flight program. In a week it was number one.

* * *

FTM 201 would launch from Cape Kennedy and fly out over the Eastern Test Range with dummy warheads coming down to target locations near Ascension Island. There, tracking systems would send data back to Autonetics where Lucille Randolph and her team would analyze the data to assess the achieved accuracy—at least, assuming the launch was a success and wasn't blown up by the range safety officer.

After many trial runs and bug fixes, we began to think MFS was ready to try for a complete mission simulation run. The Flight Weenies gave us a deck of cards representing the latest flight program for FTM 201 to be read in and executed by MFS.

It was nighttime when Larry and I transferred the MFS deck, now taking up most of three large boxes, to the Data Center. As we transferred the decks from my car to a cart inside, one of the boxes skewed sideways to overhang the edge of the cart, twisted open, and fell to the floor. Stunned, I stood for a moment in the middle of this huge mess of cards.

Larry stooped to pick up the cards. "Stop!" I commanded. "Don't touch a thing."

Although they were broken apart in small piles on the floor, the cards were actually still in order. I carefully started picking them up in groups and managed to stack them all back in the box in what I believed would be the correct order.

We submitted the job to an operator. It took a long time to get all the MSF cards, along with Job Control Language cards, loaded into the System 360 input stream. It was essential to hang around for the simulation to "get off the ground," so to speak. Often, it would fail because of a single keystroke error. This time, the simulation started to run, so we spent time sitting there, drinking coffee, and chatting. As it turned out, these late night sessions led to a deep, lifelong friendship.

"Larry, I was wondering how you got hired in to Autonetics," I said.

"Well, actually it's quite a story." Larry began to go over the world situation at the moment.

He referenced the fact that the U.S. was committed to two huge, strategic initiatives that required many scientists and engineers. The first was the Cold War competition against the Soviet Union. Both countries were developing amazing weapons systems to gain military (and therefore political) superiority over the other. The U.S. was developing and refining its strategic triad of weapons systems—ICBMs, shorter range submarine launched ballistic missiles (SLBMs), and the nuclear bomber force. These systems required tens of thousands of scientists and engineers for development programs.

Simultaneously, the United States and the Soviet Union were competing in a "Space Race," which by the late 1960s had evolved into a race to be the first to put a man on the moon. This huge initiative also required tens of thousands of engineers and scientists.

At the same time, we were involved in the god-awful war in Vietnam with its demand for draft age men. In 1968, nearly all able-bodied men were being drafted as soon as they graduated from college.

As I listened to Larry's careful explanation, his intelligence impressed me. I could tell he had told this story before. It almost seemed rehearsed. After pausing, Larry told me something I didn't already know.

"To ensure the Cold War initiatives are fully resourced, the government was allowing selected aerospace organizations to send recruiters to many universities around the country, concentrating on those having strong science and engineering programs."

"I didn't realize that," I said, "I got hired just by walking in the door."

"Really?" He went on to say, "There is a special draft deferment category, class II-A, for engineers and scientists hired to work on selected critical Cold War systems. Minutemen III is one of those critical priorities. Recruiters from Autonetics and many other companies and government labs came to Montana State in the fall of my senior year. In October of that year, Autonetics offered me a job starting in the following June. I accepted the offer in the late fall.

"The Polson, Montana draft board felt I should serve in the military. They reclassified me from II-S as a college student to I-A, available for military service, about twenty days before I graduated with a B.S. in physics. That was a little over a month ago. The only reason I didn't get drafted was because there was a thirty-day appeal process."

"What happened with the appeal?"

"Well, I appealed my reclassification and traveled back to Polson to make my case. They declined to change my I-A classification, but the thirty day appeal period provided me with just enough time to graduate and head to Anaheim where Autonetics got my Selective Service file transferred to Southern California. I must have avoided receiving a formal draft notice by five to ten days. If I had received the draft notice, Autonetics would not have been able to do anything to change or delay it."

"Wow, I am very glad it worked out okay," I said. "I was a couple years older when the big draft started, but just like you I fill out a deferment form every month anyway."

MFS ran without crashing that night. It took an hour and a half on the Model 65 System 360. The operator handed us a foot-high stack of fan-fold paper.

"Looks like we finally did it," I said to Larry. Scanning the messages on the last few pages at the top of the stack, "Looks like MFS ran to completion. The RVs flew all the way to the targets!" We loaded the MFS deck and printout into the car and took everything back to my desk.

"Let's call it a night," said Larry, and for once I agreed. Larry had been renting a room in Dick Capello's house in Orange and would head back there. "Goodnight," he said.

"Goodnight. See you tomorrow." As I drove the short distance to my apartment, my thoughts remained focused on FTM 201 and the blizzard of computations that must have been executed by MFS. After more than a year of effort, it appeared that all was in order. The simulated flight was finally successful. I slept soundly until late the next morning.

* * *

Back at my desk, I began to examine the previous night's run. Interestingly, I found the first inch and a half of output had been rising in the printer's tray faster than the simulated missile would be rising after launch. I lugged the foot-high printout over to Hlavac's desk and dropped it there with a thud. The flight programmers had been anxiously awaiting a working MFS run now for weeks. A few hours later, he came back to me.

"Hey Ralphie," he said. "The RVs overshot the test range. As near as I can tell, they impacted somewhere in a lake in Nairobi."

"Are you serious?" I knew Nairobi, Kenya, was way the hell over on the east side of Africa.

"Well, something like that. They went way long. You'd better take a look at MFS." he said. The Flight Weenies had a knack for getting my goat. They appreciated my stunned open-mouth look every time they came up with some new problem.

That night, I laid awake in bed trying to figure out what could cause MFS to fail so spectacularly? For the Nth time, I came back to the possibility of sign errors. Between the directions that a first stage nozzle would rotate based on a voltage output from the D37D to its effect on the motion of the IMU's stable platform there were about a dozen matrices and equations, any one of which could harbor a sign error. But what if a pair of sign errors counteracted each other? With counteracting sign errors, MFS would run okay without catching a possible sign error coming from the flight program. I barely slept.

The next day I was again confronted by John Hlavac, this time with a sheepish grin on his face. "You remember the flight that went long?"

"Of course. I thought about it all night."

"Well, I hate to say, it was a flight program problem. Something went wrong with calculating the velocity to be gained."

"You mean it wasn't MFS?"

"Nope. How about that? MFS seems to be working all right," he said.

I felt so good and relieved. MFS had done its job and had found its first flight program error.

* * *

"Betty Jo, you are simply going to have to put in more effort to finish VIDEO in time for FTM 201," I said, hovering over her desk. Her work was falling behind, and I had begun to resent that her efforts generally ended around 6:00 pm every day.

Looking up with a hurt expression, she said in a voice loud enough to be heard by the other cubicle residents, "Ralph, I have a family to take care of."

It was a week before the first launch of a Minuteman III and I was putting pressure on Larry and Dave as well. I thought Greg, for the most part, was pulling his weight. I had put in the bulk of the effort, writing probably as much as fifty percent of the working code. "Show me your latest VIDEO output," I commanded.

"I don't have any," she said.

"Why not?" She had been looking pretty stressed out over the past few days, and that bothered me.

"The 4020 caught on fire. Burnt up a bunch of the insides. It's going to take over a week for them to fix it."

"Are you serious?"

"Quite."

Indeed, it was the case. I called the Data Center, and they verified the bad news. With the demise of the SC4020, the Flight Weenies were going to have to make do with the mountainous printouts from MFS. If they wanted graphs, they would have to start drawing them by hand on sheets of graph paper.

A few days before the launch, Betty Jo was acting pretty strangely. She was walking around the cubicle holding up the thumb, index, and middle fingers of her right hand emulating the scrunched configuration of the PIGA coordinate frame, while singing,

"This little PIGA went to market,
"this little PIGA stayed home,
"this little PIGA had roast beef,
"this little PIGA had none,
"and this little PIGA cried 'wee wee wee'
"...all the way to the target!"

* * *

I was called into a meeting the day before launch. A dozen of us sat around the rectangular table, with a senior Air Force officer at the head and Dick Snyder at the other end. John Sweet and John Hlavac were there as well, along with other engineers and managers I didn't know. I supposed similar meetings were happening at the other Minuteman contractors all around the country.

"I think we might as well just go around the table," said the Air Force guy. Starting from his left, each person spent five minutes going over their part in FTM 201, to decide whether they would sign off on the launch. I was fifth in the row, in charge of simulating the flight of the first Minuteman III missile. I had gone over the MFS output myself with a reasonable amount of sleep-deprived care. I did notice a small anomaly occurring after second stage ignition. We had included the simulation of an accelerometer mounted on the body of the second stage that had been there for the Minuteman II vehicle, supplying angular rate data for use in the flight program control equations.

For Minuteman III, this was deemed unnecessary, and the plan, instead, was to derive the angular rate information by numerically differentiating the counts from the platform gimbal resolvers. Thus, for FTM 201, the flight program would use this method in lieu of reading data from the second stage accelerometer. I knew it was tricky to differentiate such a digital signal since the resulting angular rate would be noisy. If it succeeded, an estimated twenty million dollars could be shaved off the Minuteman III program.

"Well, there is a small transient oscillation after stage two ignition, but the flight program seems to quickly null this out," I announced. "So, I think it will be all right."

"Can you tell me that your simulation is correct?" asked the officer.

"All I can say is we have done our best, sir." I was asked this question many times in the weeks and months afterward, and my answer was always the same. He thought about this for a moment before going on to the next person sitting on my left. In turn, the two Johns affirmed all was ready. As the last man was queried, it was agreed that the launch should proceed.

A special Mylar eight-channel tape, silver on one side and a beautiful blue on the other side, containing the finished flight program had been punched out and carried by plane to Florida by an Autonetics engineer. This tape, now considered a mere piece of hardware, would be read into the missile in preparation for launch.

* * *

I awoke with a start, reached over the side of my bed and shut off the alarm. I had gone home early the night of the sign-off meeting, since there was nothing much else to do. Sleep had come easily and now I was refreshed for the scheduled 4:30 pm launch.

"Where have you been?" asked Greg, excitement in his voice. It was early afternoon and I had arrived at my desk.

"The launch went perfectly," said Larry.

Still a bit groggy from sleeping fourteen hours straight, I said, "What? The launch isn't scheduled until 4:30."

"That was 4:30 Eastern Daylight Time," said Greg. "1:30 pm here at Autonetics. You just missed it."

"Yeah, everyone went to the auditorium building. It was packed, so we stood just outside the door. They called out the events as they occurred, stage 1 ignition, stage 2 ignition, and so forth all the way to Post Boost," said Larry.

"It was a complete success. The RVs landed smack on their targets." said Greg.

I didn't have much to say. After working like hell on it for a year and a half, I had slept through the launch of the first Minuteman III missile.

"Congratulations, everyone. That was quite a show." It was a jubilant, smiling John Sweet, peering into our cubicle. "Beer all around at 5:00," said John.

I arrived at the Go-Go place shortly after five. Altogether, ten or twelve of us gathered there, including the MFS team, the Flight Weenies, Alan Bernstein, and a few others. Greg had grabbed a full-page black-and-white pre-print advertisement handout for the LA Times showing a dramatic photo of the missile leaving the launch pad trailing fire and smoke. Beside the photo, it read in large type, "Friday, August 16, at 4:30 PM EDT, we helped make the nation even more secure. It's a good feeling."

Four days earlier, I had picked up a copy of the Times with the headline blazing across the page in two-inch high letters "NEW U.S. WEAPON, To Show Multiple-Warhead Missile," and smaller column header by George C. Wilson, "Seen Having Heavy Impact on Arms Race."

"WASHINGTON — The United States this week will show a new weapon which will have an incalculable effect on the world's arms race.

"It could dampen it — and thus fulfill President Johnson's hopes — or escalate it and confirm the fears of many scientists.

"The weapon is called MIRV, an acronym for Multiple Independently Targetable Reentry Vehicles. It is several nuclear genies in one bottle in the form of several H-bombs atop one missile. The deadly package of H-bombs is slated to be carried by the Minuteman 3 ICBM and the Poseidon missile. The Minuteman 3 will be put in underground silos while the Poseidon will replace the Polaris missile in our nuclear submarines."

The article continued on an inside page, where it discussed strategies for using MIRV in a nuclear war. Opposing views were presented as to whether it was a good thing or a bad thing.

None of us were concerned about any of this as pitchers of beer were passed around the adjoining tables. We were all ecstatic that the damn thing actually worked. Sharing in such a monumental achievement served to cement my deep friendship with Greg and Larry. Likewise, with John Hlavac and Alan Bernstein.

By the time we staggered out of there, we had finished a thirteenth pitcher and the topless Vietnamese dancer had stuffed quite a few dollar bills in her G-string.

Four days later, the Soviet Union invaded Czechoslovakia.

25 Seeking Celestial North

W E WERE ALL BASKING in the glow of the successful first flight when several flight program errors were discovered while examining the post-flight telemetry data. Thankfully, the errors were all relatively minor in terms of their effect on overall performance. There was a small error in the TRW-supplied targeting data, an incorrect sign in the low-order term of a flight program control equation, and a longer than desired lag in controlling the vehicle's roll.

Given these problems, it behooved the Flight Weenies to re-examine the MFS output for FTM 201. The flight program anomalies were there in the printouts and would have been readily apparent by looking at graphical results from Betty Jo's VIDEO program, had the SC4020 not caught fire. But hindsight is always perfect.

With a little bit of breathing room until the next test flight, I started spending my time trying to document our work, carefully drawing pictures of the various coordinate frames, views of the missile, and flow charts illustrating the sequence of various calculations. I wanted the legacy of Minuteman Flight Simulator to endure.

I needed to understand everything going into MFS, particularly, the calling hierarchy. Greg had suggested I start drawing a "tree diagram" on large sheets of paper to keep track of who is calling whom. Small circles represented individual routines and arrows pointed to each called routine, not unlike the branches on a tree. If a particular routine called, say, three other routines, there were three arrows emanating from the higher-level routine, whose arrowheads terminated on each of the routines it could call.

Soon, I came to realize my complete tree diagram would cover more than a hundred square feet of paper! Instead, I came up with a solution that I called a "pseudo-tree" diagram. Circles represented individual routines as before, but instead of connecting the circles with arrows, they were connected by what looked like railroad tracks. Each modular routine had only one track leading into it at the top of its circle, and only one track leaving it at the bottom that branched out with curved connections to all the routines it could call. Eventually, there would be about one-hundred fifty circles on a single large sheet of paper, all connected with tracks, proudly pinned on the wall of our cubicle.

Of course, I also struggled to keep track of the variable data relationships among all the routines. Most of these variables were stored in FORTRAN COMMON blocks accessible to many different routines. I made diagrams showing how each main routine was executed in sequence within the integration loop formed by the calls to Runge and Kutta. I added curved arrows labelled with important variable quantities, like thrust, drag, and the relevant coordinate frame transformation matrices. I worked with these diagrams until I was sure I had an intuitive "gut" feel for how everything was working. In my earlier years as an amateur artist, I had painted large abstract creations in tempera as wall decoration. Now, as I gazed at the newly created diagram of the boost-phase vehicle dynamics, I thought it possessed a similar beauty.

* * *

Greg had surgery for his bum knee that summer. While in the hospital, he had a lot of time to think about the future.

"Ralph, I've decided to leave Autonetics," Greg told me.

"I don't understand. We're working well together," I said. I knew I would miss Greg, but I was sure our friendship would somehow survive.

"My alma mater is starting a new Ph.D. program in computer science. I've decided to join it so I can maybe get some more smarts." Greg left Autonetics and joined the new program in September.

I missed working with Greg, but flight testing continued during the fall of 1968, with flight program refinements and subsequent testing runs with MFS. The second test of Minuteman III, FTM 202, was launched from Florida on October 24. The mission was a complete success.

* * *

The TWA Boeing 707 pitched violently as we crested the Rocky Mountains heading for Kansas City. Some of the passengers around me conversed fearfully. From my left window seat, I watched as the wing bounced up and down, taking the loads. It was my first trip on a jet, headed back at Christmas to visit my daughters and parents. We had left LAX in the afternoon, and it was now turning dark.

As we approached the KC airport, it began to snow. We were flying in nighttime blizzard conditions, bouncing violently as the flaps were deployed on the final glide slope. Suddenly, the pilot applied full power with a tremendous roar as we started to regain altitude. He came on the speaker: "It's too icy to land. The plane in front of us almost slid into the river!"

We circled around for a half hour and once again started our approach for landing. People fidgeted. Two rows in front of me, a gray-haired man with a large nose took a harmonica out of his pocket and began to play softly.

The plane touched down and rolled to a stop in a large field of snow with no buildings or other structures in sight.

After an hour, trucks began to approach us, then several buses, and a set of stairs was pushed up to the front door. All of us made it down the stairs and into the warm buses. They took us back to the KC airport, arriving there at three o'clock in the morning. I learned later that we had landed on a runway fifteen miles northwest of Kansas City. It was the site of the new Mid-Continent International airport, which would not be completed for another three and a half years.

Later in the morning, another short uneventful flight on a prop plane took me to Wichita.

My father had driven to Manhattan to pick up my girls, bringing them to Wichita. Carol was about to turn seven, and Dana had turned five. I was pleased they were wearing the brown and green winter coats I had bought for them. The storm had brought a half foot of snow, so we had a great time playing in the yard and making a big snowman. It was wonderful to be with my parents and siblings as well.

"Dad, I want to tell you about my work at Autonetics," I said, excited to tell my father at last about the Minuteman III, MFS, and the first two successful launches.

"How long will this take?" asked my father.

"Not long." I could tell he was not eager to be pinned down and subjected to one of my lectures. "I worked on this new kind of missile with up to three warheads," I said. "It will cause the Russians to think twice before attacking us with their missiles." I showed him the LA Times front page.

"That's good, Ralph."

"I wrote a program that simulates the flight of the missile," I said. "You see, the missile—Minuteman III—has a small computer on board that flies the missile. My program runs inside a big IBM computer and pretends to be the missile, including the small computer on board. We call it 'simulation.' That way, we can test out the program that'll fly the missile without launching a real missile."

"I thought you told me they already launched real missiles," he said.

"Yes, but not before the program is checked out using the program I wrote." I was proud of the work I had done, and he could tell I was. But understanding even a little of my work was beyond him. I realized there was no way he could comprehend what I was about to tell him.

"So, what are you going to do now?" he asked, quick to change the subject.

"I'm not sure. I am thinking of maybe leaving Autonetics. Maybe I will try to go back to school." I said.

"Back to K-State?"

"Probably not, I don't know what I will do."

"Have you thought about coming back to Wichita and becoming a finish carpenter? You've got the skills. Finish carpenters make good money and they only have to work inside, out of the weather."

All of us gathered around the TV set in the evening of December 21 to watch the liftoff of Apollo 8. After the boost from the Saturn V S-IVB stage, Frank Borman, James Lovell, and William Anders were headed for the moon. The family gathered to listen on Christmas Eve, as each of the crew, now in orbit around the moon, read passages of the Biblical creation story from the Book of Genesis. It meant a lot to my father.

* * *

In the New Year, back at Autonetics, we were all working toward FTM 203, the third flight of a Minuteman III. Additional flight program event messages had been added to MFS, as well as some model refinement and inclusion of Larry's eighth-order gravity model. Several more flight program routines had been added for more aggressive PBV maneuvers to prevent the axial engine plume from striking the chaff cloud.

Betty Jo answered the phone on her desk, stood up, and looked back toward me with a terrible expression.

"Is everything all right?" I asked.

"My sister back in Kansas just died," she said.

"What happened?"

"She got the Hong Kong flu a few days ago and passed away this morning."

I was stunned by this news and stood there helplessly. She put on her coat and headed out. Larry had overheard. "Maybe you should have offered to drive her home," he suggested.

Betty Jo had already left the building when I caught up to her. She thanked me, told me it was OK, and said she could drive home herself.

I was sorry for Betty Jo and was concerned with the safety of my own family back in Kansas. Later, I learned 34,000 people in the United States had died from the Hong Kong flu pandemic in 1968-9.

* * *

I picked up Georgie before dawn for what was our last drive southward. As we drove, she talked seriously about our relationship. I had just returned from my trip to Kansas.

"I'm convinced you will get back with your wife," she said.

"My marriage is over. Do you want to have sex with me?" I asked. "Is that the problem?"

"No, that's not it," she said. "You know, we really don't have much in common. I just know I could never fall in love with you." During the long drive back to her apartment, we decided to end our relationship. It was finished. I knew I had resisted getting involved too deeply with anyone, but I was sad that another dating relationship with a woman was over.

* * *

FTM 203 launched on March 26, 1969 from a new silo near the one used for FTM 202, flying out over the Eastern Test Range. After its third stage ports were blown out terminating thrust, the Post Boost Vehicle should have taken over. Something went wrong causing it to tumble like a mindless dead weight in a long arc, plunging into the Atlantic 4000 miles downrange.

Eventually, it was determined by analyzing telemetry data that a short circuit had occurred in the D37D, likely as a result of the mechanical shock from the third stage thrust termination event. Mercifully, neither the flight program nor MFS was at fault.

All of us worked night and day for the next two launches, FTM 301 and FTM 204. The former was to be the first Minuteman III launch from Vandenberg Air Force Base in California. The flight programs were still being revised, requiring quite a few more MFS runs.

Between simulation runs I had some time to study what a real Minuteman missile would be doing in its silo before launch. Another program, the

Minuteman III Ground Program, would be running in the D37D computer inside the Post-Boost Control System. Alan Bernstein of the Bernstein-Hlavac duo was an architect of the ground program. While the missile rests in its Launch Facility (LF), a twelve-foot diameter steel-lined reinforced concrete launch tube (silo), this program communicates with the Launch Control Facility (LCF) ten miles away using a Hardened Intersite Cable System (HICS) and secure radio channel. There, two specially trained missilemen can go through pre-launch preparations. Each LCF communicates with a field of ten widely dispersed LFs, each with a Minuteman missile.

Automated functions of the ground program operate continuously day in and day out for months on end. Some routines wiggle the first stage nozzles back and forth once in a while to make sure they are not stuck and move in the correct directions. Other functions monitor the environmental conditions in the silo and the site security. But most of these functions involve the inertial measuring unit (IMU), the *heart* of the missile.

I knew the triple-gimballed IMU platform is stabilized by a pair of ultra reliable two-axis free-rotor gyroscopes supported on hydrogen gas bearings. One gyro serves as the pitch and roll axis stabilization reference; the other provides an azimuth stabilization reference (the remaining axis is electrically caged to prevent movement). The self-generated gas bearings have some wear associated with startup and shutdown. During startup, a much higher than normal voltage is applied to accelerate the gyro rotors quickly, generating a cushion of gas. The displacements of the spinning rotors are precisely measured using electrical capacitance sensors.

In addition to the stable platform whose gimbals are continually torqued to match the Earth's rotation rate, and the three pendulous integrating gyro accelerometers (PIGAs), the platform carries a precision *gyrocompass*, four *electrolytic level sensors* and two precision *alignment blocks* with mirror surfaces.

I learned more about the PIGAs. Each PIGA is a sealed cylindrical tube one point six inches in diameter containing a spinning gyroscopic rotor. The precision rotor is suspended by a hydrogen gas bearing on a short cantilever at right angles to the nominal direction of flight. The unbalanced torque of the spinning rotor would cause gyroscopic precession to occur around the nominal direction of flight, except a torque motor prevents the precession from occurring while the missile is on the ground. This torque, derived from a D37D voltage output, is proportional to the acceleration of gravity at the launch site.

As Betty Jo had demonstrated to us, the three little PIGAs on the IMU's stable platform are arranged so they are along the axes of a "scrunched" coordinate frame clustered around the nominal direction of flight. Every so often, the ground program can perform a calibration of each PIGA by rotating the stable platform until the selected PIGA points straight up, along the launch site local gravity direction. This procedure enables the bias, scale factor, and sensitive axis of each PIGA to be measured and recorded for use by the flight program. (The local gravity is independently measured once and for all using a precision gravimeter.)

The question of what direction is actually "straight up" is solved by the electrolytic level sensors located on the top, bottom, and sides of the precision alignment blocks attached to the upper and lower surfaces of the stable platform.

The level sensors themselves are slightly curved glass domes, like two-axis versions of a carpenter's bubble level. Instead of containing water, however, they are filled with a conducting liquid except for a vapor bubble. Four electrodes are arranged at ninety-degree intervals around the edges of the glass dome.

Torquers on the gimbal axes are used to tilt the stable platform, rotating it into level. Electrical circuits measure the differences in electrical resistance between the electrodes. When the differences are zero, the bubble is in the center of the glass and the sensor is level.

When the missile takes off, the PIGAs experience an acceleration additional to the launch site gravity. The computer, now transitioned to the flight program, takes over the same torque values that were maintained by the ground program. This causes the PIGA rotors to precess at a rate proportional only to the added acceleration of the missile. The angle the rotor precesses, measured by a precision resolver, is proportional to the missile's velocity.

The flight program needs precise knowledge of velocity to accurately deploy the RVs and other objects. This requires accurate PIGA calibration. The missile's position in the Launch Centered Earth-Fixed (LCEF) frame is given by numerically integrating the PIGA velocities.

Development of the intricate electromechanical PIGAs, invented in Germany by Fritz Mueller for the V-2 rocket, combined with precision gyro technology by the MIT Charles Stark Draper Laboratory, refined by engineers at Autonetics, and subsequently manufactured in quantity by the AC Spark Plug, Bendix, and Honeywell companies for Minuteman has taken many years fraught with difficulty. For better or for worse, the PIGA is among mankind's highest achievements.

Precise knowledge of velocity is one thing, but hitting the targets also requires precise knowledge of the IMU's platform azimuth at launch. The azimuth is the angle between the platform (P) X axis and the direction of true North. The gyrocompass assembly at the top of the stable platform has another small, gimballed gas-bearing-supported gyro that nominally would remain at a fixed orientation while the Earth rotates beneath it.

But instead, the gyro's rotation axis is continually torqued to remain horizontal. This constraint causes the gyro's rotation axis to precess in the horizontal plane as the Earth rotates, until the axis points toward Celestial North, after which the axis direction remains fixed with respect to the platform.

Celestial North is the direction straight up from the Earth's North Pole. To an observer standing at the exact North Pole of the Earth, the Celestial Pole would appear directly overhead, presently only around three quarters of a degree from our friend, the relatively bright star Polaris. In another thousand years, due to the Earth's own precession, it will be near the orange giant star Errai. Hopefully, by then, Minuteman missiles will no longer be necessary.

In reality, the gyrocompass assembly has many unavoidable small manufacturing errors, unanticipated frictions, small mass imbalances and the like which lead to small errors in the pointing axis. These errors are largely corrected by a special gyrocompass calibration procedure performed by the Minuteman III ground program after startup, and every ninety days thereafter. This procedure, called the Self Alignment Technique (SAT), rotates the stable platform between upright, inverted, East-, and West-facing orientations, using the precision level detectors for reference.

There is an optical "autocollimator" device on the wall of the silo, with small glass windows in the side of the PBCS and the IMU itself allowing optical access to the mirror surfaces of the alignment blocks. The autocollimator is a high precision angle measurement instrument which can measure angular deviations with accuracy down to fractions of an arc-second. It works by projecting an image onto one or another of the alignment block mirrors and measuring the deflection of the returned image by means of an optoelectronic detector. This deflection is communicated to the IMU through an umbilical cable and used as a primary reference for the self calibration technique.

By periodically turning the stable platform back and forth through the four cardinal orientations, and filtering the autocollimator readings, the major errors cancel out. This procedure also eliminates the need for accurate alignment of the autocollimator in the silo with respect to a precise surveyed monument located near the silo. SAT provides a more accurate determination of azimuth than was previously possible with the Minuteman II IMU. It all boils down to how accurately Celestial North can be found.

In total, the *heart* of the missile has two large gyros to stabilize the platform, three smaller gyros in the three PIGAs to measure acceleration and velocity, and another gyro in the gyrocompass to determine Celestial North. Add to this, the D37D *brain* of the missile has the rotating mass of its memory disk. Thus, a total of seven masses are spinning at high speed on hydrogen gas bearings, day in and day out without failing for many months: a remarkable achievement.

J. M. Wuerth, a senior Autonetics manager, determined that aligning the IMU platform in Minuteman III to enable hitting the targets is like threading a needle from 400 feet away, or kicking a field goal from 50 miles away. Of course, it also depends on knowing the precise positions of the launch site as well as the targets.

There is no need for a Minuteman Ground Simulator, since the ground program can be checked out using the real hardware. I was relieved that MFS only needed to simulate the missile beginning at first stage ignition.

* * *

Preparing for FTM 204 was fairly routine. Betty Jo's VIDEO program was working well, with graphical output from the repaired SC-4020. The launch was scheduled for April 22, from the same silo as the one used for FTM 203. It would fly out over the Eastern Test Range with impacts near Ascension Island.

FTM 301 would be another case entirely. The missile would launch from Vandenberg Launch Facility 02, located at 120.5846 degrees West Longitude and 34.8448 degrees North Latitude, scheduled for the night of April 11, 1969. It would fly out over the Western Test Range, with dummy RVs impacting near Kwajalein Atoll in the Marshall Islands. A sign error might cause it to fly eastward over the continental United States, instead.

Vandenberg Air Force Base is a three-hour drive from Anaheim. A group of top managers had been selected to travel there for the launch. I asked if I could go, and the answer was emphatically "No."

After working so hard on MFS, having seen only a sketch of Minuteman III, I was determined to find a way to get there and see a real one in person. The afternoon of April 11, I drove to a map store in Orange, and purchased a high-resolution United States Geological Survey map of the part of Vandenberg AFB showing the location of Launch Facility 02. After returning to Autonetics, I studied the map carefully to find possible access points that would take me near the launch site.

"Larry, I'm driving to Vandenberg tonight to see the launch. Do you want to go with me?"

"Well, yes. Okay," answered Larry.

"Can I go too?" Betty Jo had overheard us.

"We're going in my MGB. There's only room for two," I said. "Besides, there could be some strenuous hiking involved." I knew this could be one of the most daring and thrilling adventures of my life.

26 Launch!

NIGHTTIME IS UPON US as we drive for nearly four hours northwest toward our objective in pleasant spring weather, light traffic impeding us on the 101, as I pilot the MGB through Santa Barbara. Larry calls out smaller roads as he trains my flashlight on the USGS map. At last, we find the small dirt road. We arrive at a little-used gate of Vandenberg, backing the car into the bushes.

Soon, a vehicle approaches the gate from the other side. A jeep with two guards stops ten feet from the gate, the beam from its headlights reaching out to us. After a minute, it turns around and heads back.

I feel my heart beating. "A close one," I whisper.

"Are we actually going to do this?" Larry asks.

"Let's go. I think we'll be all right." Stepping out of the car, we cautiously click the doors shut and accommodate to the darkness and warm pungency of the eucalyptus around us. It is quiet except for sounds of insects. Twenty feet to the right through the weeds, we reach a three-strand barbed wire fence. Larry reaches down and spreads the wires apart, and I slip through. Then, I do the same for him.

We head into the grassy field, hearing faint sounds of what must be cattle. We make our way uphill through darkness for a quarter mile. Suddenly, I lose footing. We both stop. "I'm going to turn on the flashlight. This is insane. We could fall into a hole."

"Don't do it. Somebody will spot us."

I point the flashlight ahead of us and push the button for a few seconds, revealing a grassy incline for the next twenty steps, then blackness. As we slowly make our way up to the top of the rise, we see a ravine, looking to be thirty feet deep. "It's not bad. I'll keep the light pointed downward and we can make it down and up the other side." My stomach starts to tie itself in knots, thrilled to hopefully witness firsthand the results of our effort. On the other hand, I am terrified there might be some hidden flaw in the code causing everything to explode in a ball of fire.

The lights of Santa Maria appear to the north. We spot the fence in the starlight, the neat, straight line heading up the mountain, marking the base boundary. "We should stay at least ten feet away in case there are sensors that might detect us." Now, gaining elevation, we can barely make out the western edge of the continent. As we crest a ridge, we spot below us the unmistakable square pattern of red lights marking the outline of Launch Facility 02.

FTM 301, in its silo, is only a scant 400 yards away. Its stable platform is precisely level, aligned by the ground program's Self Alignment Technique, being continually torqued to match the Earth's rotation. Likewise, the three PIGA gyros are being continually torqued to prevent them from precessing. All is ready for the flight program to take over and issue the stage one ignition discrete.

In the far Pacific, a team of engineers are operating precision radar systems that will track the incoming dummy warheads. We wait silently in the darkness for the countdown and launch. We know the missile is poised to leap out of its silo. Presently we hear talking over a loudspeaker. I check my wristwatch—only fifteen minutes until launch. It is nearing midnight. A waning crescent moon is rising in the east. Someone calls out parameters and numbers related to the missile. Silence. I breathe deeply, sweating from the climb. My stomach tightens some more. Silence. Without warning, a loudspeaker calls out "ten, nine, eight, seven, six, five, four, three, two, one, zero!"

Suddenly, an explosive column of white-hot fire roars out of the silo, instantly turning our hillside into broad daylight. We instinctively hit the ground as the heat strikes us. The bright flare blossoms in front of my eyes as I reflexively gasp! The pressure wave and sound hits us in the chest four seconds later. I feel the ragged vibrations of the rocket lifting off as a tingling through my trunk, arms and legs. I squint to tolerate the brightness of the light. The sound hurts my ears as the rocket soars upward.

"Oh my god!" I say, as I breathed a sigh of exhilaration.

"Holy shit," comes from Larry.

The missile rises swiftly out of the hole and pitches over toward the targets, pointing its thunderous exhaust almost directly toward us. "Skirt jettison now," I say, although we couldn't see anything coming off the missile. From MFS, I know exactly what is supposed to happen. A few seconds later, I call out "First roll maneuver!" The missile is only a bright star now as it heads out over the Pacific. Thirty seconds later, there is a momentary bright flash.

"Stage 2 ignition!" says Larry.

A bit later, I know the second stage skirt is jettisoned. A minute later is a smaller flash. "Third stage ignition." In another minute the star winks out. "Post boost," I say.

* * *

We tracked the missile with our own eyes from takeoff to a point in space over two hundred miles away in a bit more than three minutes, witnessing the launch of a Minuteman III from a vantage point likely as close as anyone has ever been. In another fifteen minutes, if all goes well, the three dummy RVs should be streaking out of the Kwajalein sky toward their targets.

"Well, what do you think?" I ask. It's the culmination of years of effort going all the way back to my White Vulcan rocket.

"Awesome!"

What if the RVs were real nuclear weapons directed at an enemy? One warhead could knock out a missile silo, another could destroy an airfield, and a third could destroy a nearby medium-size city. Millions of people would die.

As we start back down the hill, the jeep, far below us, once again approaches the gate near where my MGB is hidden. It pauses for a long five minutes as we wait in fear of being discovered. The jeep turns around and drives back toward the lighted area of the base.

"We're in the clear. They didn't see my car," I say. Finally, one of my rockets actually seemed to work.

27 Endings and a New Beginning

T HE NEXT DAY at Autonetics, John Sweet and Dick Snyder were smiling and jubilant as they announced their impressions of FTM 301 to the Flight Weenies. "The launch was spectacular! Just before midnight!" said a happy-looking John Sweet. "We had a great view from the blockhouse."

I smiled at Larry. "Yeah, we know," I said quietly, so no one else could hear. Telemetry data showed that the flight of FTM 301 was a complete success. It performed exactly as MFS had said it would. The missile reached an apogee of eight hundred miles, and the RVs hit near their target locations five thousand miles downrange. I thought of all my amateur years trying to design and build rockets, and my efforts to simulate their flights using the IBM 610 and Bendix G-15 computers. This felt pretty good.

We ran some new simulations for FTM 204, which would launch from Florida only eleven days later, and worked in a few more user interface requests from the Flight Weenies. The missile launched on time and was completely successful.

FTM 205 flew successfully on May 27. But on May 29, FTM 302, the second Minuteman III to fly out of Vandenberg, failed to enter the post-boost phase. From telemetry data, the fault was traced to particle shorts in the D37D. Nothing to do with software.

Programming tasks on MFS were getting fewer, allowing me to spend more time outside of Autonetics. I dropped down to working only fifty to sixty hours a week. By now, we had the use of two IBM 360 Model 65s, and frequently used both at the same time. The Autonetics Data Center had become the largest computer center in the state of California. It was getting hard to find a parking space, since the employee population had climbed to about thirty-six thousand people.

The repetitive cadence of making MFS runs at the Data Center was starting to get boring. With more time on my hands, I longed for the companionship of a woman.

* * *

"When is the last time you've been with a woman?" she asked. I had noticed the dark-haired young waitress a few times at the burger place near my apartment. I usually sat in the back, and that afternoon she was making her way toward me as she cleaned the tables and chairs. The popular song, *"Working My Way Back to You"* popped into my head, so I mentioned it to her when she reached my table, and she laughed. I had asked her if we could talk after her 4:00 shift was over, and she said yes.

"What do you mean?" I replied. We had wound up walking the short distance to my apartment and were now sitting next to each other on my couch.

"I mean, you know—*been* with a woman?" she asked again.

"Well, I've had a couple of girl friends the past several years since I moved to California after my wife left me."

"Okay. What I mean is how long has it been since you've had sex with a woman?"

I could feel a flush rising in my face. *How can this girl I had met only an hour ago ask such a question?* I stammered once or twice before coming up with an answer. "I guess it's been about two years."

"Oh, you poor guy. I feel really sorry for you," she said.

"It's okay, I'm fine," I said, averting my face.

"I can help you. I want to make love to you."

"You do?" I asked, incredulously. "Now?"

"Why not?" she replied.

My mind rushed ahead of me—knowing this could be the stupidest and most dangerous thing I could ever do; knowing there would be not even a hint of love here; not knowing anything about this girl, who she was, where she came from; not knowing what she might want to do with me; thinking it would betray my relationship with Audine; knowing the love I had for Audine was long gone;

thinking about my lost love for Frankie and even for Georgie; knowing she might even be married; knowing I could get a venereal disease; knowing she could get pregnant; knowing this was a bad idea; knowing how I had laid awake at night aching to have someone with me. Knowing all that, looking directly in her trusting brown eyes, I simply said, "Yes."

Afterwards, lying on my bed, she said, "Thank you. That felt good."

"Me too," I said. *It feels wonderful; simply wonderful!*

"I'm sorry, but I have to go to my night job now."

"Night job?"

"I also work in the new Kotex factory in Fullerton," she said. "From now on, we can only do this at night, after the midnight shift ends. I will just come by and knock on your door."

True to her word, she stopped by every week for a month. At one of our late-night meetings she said, "This is our last time. I'm moving to San Francisco in the morning." She left quietly as I was half-sleeping; heard the click as she closed my door. I never saw her again.

* * *

I desperately wanted a relationship. I needed the company and pleasure of a woman; the physical relationship with a woman, but I had no idea where I could meet someone. I was not a drinker, so I was leery of picking up someone in a bar, and I was deathly afraid of drug scenes and Wesson Oil parties.

Sometimes, in the evenings, I would get my dinner at H. Salt Esquire Fish and Chips, or at a Sizzlers restaurant. The Sizzlers in Anaheim had good steaks. A cute and pretty girl—looking to be a few years younger than I—worked there most evenings. I would spend time thinking about her during my meal while stealing furtive glances toward her. The person you ordered from at the counter would also deliver it to you at your table. Several times I would shift a person or two back in line in the hopes I would get to order from her.

I'm in luck! I tried to drink in the vision of her soft blue eyes as I placed my order, knowing she would be the one who would take it to me fifteen minutes later. Now, at my table, I had to think quickly—summon up my courage to come up with something to say to her. *Here she is!* I looked up as she put the tray down in front of me.

"Hello, what's your name?" I asked.

Hesitating for a second, she asked, "Why do you want to know?"

"I..., I just wanted to know if we could go out sometime."

"I already have a boyfriend, so the answer is no," she said plainly, turning back to the kitchen. But our eyes had met briefly, telling me I perhaps still had a chance.

Back at Sizzlers the following week, I ordered my steak and fries from a woman in her fifties. The girl of my dreams was still there, waiting on other customers.

Suddenly, while concentrating on my steak, a grizzle-faced man of maybe forty grabbed the chair next to me and sat down on it tough-guy style with its back toward me. Leaning in close to my face, he said quietly, "I'm her boyfriend," apparently referring to my dream girl. "You know, I used to live down in Mississippi. A n****r was bothering my friend's girl. We caught him, and I helped hold the black sonofabitch down while we cut off the tip of his dick." The man got up, returning to the kitchen. My appetite got up and left me. I never again returned to that particular Sizzlers.

* * *

I needed to get away from computer programming and back to doing physics. My experience back in graduate school had opened my eyes to the beauty and wonder of physics. I was good at computer programming, but that now seemed more related to mere engineering. Autonetics had a Research Division in another building, so I applied to be transferred there. One of the scientists showed me an interesting sensor device he was developing, a so-called "lateral effect position sensing photodiode."

I could draw on its surface with a laser beam, and the path would appear on a television monitor. I thought that was neat and told the man I would very much like to have a position in the Research Division. He took great offense — schooling me never to ask for something as arrogant as a "position" in a job interview.

I interviewed for a job at Hughes Aerospace, careful to eschew the word "position." The interviewer told me about five available areas of research, asking which one appealed to me the most. I picked the only one I felt I could sink my teeth into. He said, "Oh wait. You only have a master's degree? Sorry, we need to hire a Ph.D. for that one."

The Minuteman Flight Simulator was in good hands with Larry, so I started taking some time to sit in on a physics class at U. C. Irvine. I sat in the last row in a class on solid state physics, listening in. The professor never caught on to the fact that I wasn't a paying student.

Greg Hopwood was busy pursuing his Ph.D. in computer science at Irvine, and we remained good friends. I made many trips with him to his parents' home in Tustin, enjoying spending time with his younger siblings. Mike was a kindergartner. Chris showed me his falcon. Greg's mom, Marjorie, was like a second mother to me. Greg belonged to a club of car enthusiasts and had an AMX muscle car he entered in races. I joined in on one event with my MGB and his sister Vicki.

* * *

I decided to apply to a physics doctoral program to find out if I would even stand a chance. I worried, however, that most of the physics I had learned back at K-State would have been forgotten by now. Larry suggested I might like it at the University of Colorado. His undergraduate advisor was from there. I thought about the adventurous mountain environment and proximity to Wichita, so I applied.

Meanwhile, Greg and I started an evening adult education course in machining at a local high school. I needed to design and build an enclosure for my computer project. Before we could even touch a machine, we had to take lessons in elementary shop math. We received permission to use the tools right away, after successfully arguing that we already knew all the math the instructor was likely to teach.

"Greg, what's wrong?" I asked. He met me there with a sad expression on his face. It was June 24, 1969 and we had been working in the shop class for a month.

"It's my brother, Chris. He was camping with a friend up on Mount San Jacinto. He slipped and fell fifteen hundred feet and was killed." I didn't know what to say, so I didn't say much of anything. "Dad and I went up there yesterday and helped to recover the body."

The funeral was a few days later. There was music; one of Chris's classmates sang and played a guitar. I tried to console Greg and the members of his family. Greg's girlfriend, Marsha—a fellow student at Irvine—was there.

A month later, Apollo 11 landed on the moon. I was with Alan Bernstein and John Hlavac, watching it happen on John's black and white TV. The next day, the Autonetics Skywriter interviewed people about the moon landing: Ralph Hollis, Jr., NSD Systems and Reliability Engineering, called it *"The most momentous event ever undertaken by man, and I believe that eventually it will benefit all mankind, helping us solve our social and human problems."* A week or so later, I drove to Downey to view an amazing sight: the charred Apollo 11 command module, fresh from its trip back from the moon. The Apollo program excited the world's imagination. In contrast, it seemed that Minuteman III was just a Cold War footnote almost no one cared about.

* * *

It was late summer when I fired up the MGB and drove to the University of Colorado in Boulder. To my surprise, I had received an acceptance letter a few weeks earlier. At the University, I met with some of the physics professors, and scanned the local Daily Camera paper for rental apartments. Boulder is a wondrous place—the backdrop of the Flatirons rock formations to the west of campus is an awesome presence. I added up the cost of tuition, rent, and food— it was all more than I could afford. In four or five months it would total out my meager savings.

I spent the weekend exploring the campus and the town. I tried to remember what it was like to take physics courses—still haunted by the F midterm exam score I got back at Kansas State. I wondered if I really had what it takes to do this. *To Be or Not to Be.* In the end, I concluded that it was not to be. Reluctantly, I drove the thousand-some miles back to Anaheim.

"I thought you said you were leaving Autonetics to get your Ph.D.?" It was Al Sheue.

"I guess not," I replied. "Maybe I will, someday."

I began wandering around the building, peeking into some manufacturing activity on the first floor and stealing scraps of Teflon-covered hookup wire from the trash bins for my tabletop computer project, grateful that I still had a job.

I was sick of Autonetics. I didn't want to just make endless runs of MFS. I began to mull over my remaining options. At twenty-four, I was starting to feel middle-aged. I had failed as a husband, felt detached from my daughters, and couldn't find love. After working so hard on MFS and feeling successful, I now felt like a failure. I didn't want to go back to Wichita, to work as a finish carpenter. The dream of getting the Ph.D. in physics had receded a thousand miles away.

* * *

After the last few episodes with women, I was disillusioned with waitresses and began to consider other girls at work. For some time now, I had been chatting with Becky in the building 235 keypunch room. Petite, sweet and perky, laughing easily at my clumsy advances. Even so, I didn't know much about her, so I asked her if we could go somewhere to talk. We got to know each other a bit better then, and I learned she was the single mother of a young son. We agreed to date.

Before long, Becky was my steady girlfriend, sharing a satisfying relationship. We mostly met in her apartment, since mine was so messy with bits of the computer project lying around. We took drives around the local area and went out to eat in a few inexpensive restaurants. She wanted to have more exciting dates, so I took her to a concert featuring Roger Miller. We sat in the front row when he sang his signature song, *"Dang me."* She wanted to marry me. I wasn't sure I loved her; wasn't sure I knew what love was. But eventually, we were engaged, and I bought her a diamond ring. I met her parents who lived nearby. Her father was a "money doctor" who bought ailing companies, restoring them back to profitability. We enjoyed their nice pool and sauna.

Eager for her to meet my parents and siblings, we flew to Wichita where my little daughters appeared. Dad had flown the Tri-Pacer to Manhattan and picked them up. But after two days with my family, Becky had had enough; she flew back to California on her own. I didn't understand her reasons, but I realized our relationship was over. There was no discussion, and for a change, I felt no rejection. In fact, I felt relieved.

I returned to California with Carol and Dana sitting beside me, their first flight in a jet plane. We had a grand time together, visiting Knott's Berry Farm and Disneyland. I took lots of pictures; pasted them in a photo album and sent the album back to my girls and Audine.

The next week, Becky returned the ring, so I went back to the jewelry store to get my five hundred dollars back (one sixteenth of my annual salary), explaining that that we were no longer in love. "It doesn't work that way, son," said the kindly middle-aged woman behind the counter.

* * *

The piercing scream startled us, chilling me to the bone. "Stop! Listen..." I said quietly to the girl walking next to me. We held still, holding on to each other in the cold darkness.

"What is it?" she asked. Fear crawled up my back.

Another primal scream; a hundred yards ahead of us. "Mountain lion," I said. But I wasn't sure—I had never heard such a terrible sound. Maryann and I had decided to hike to Ostrander Lake in Yosemite National Park. We planned to camp there overnight, returning three miles in the morning back to my MGB parked off Glacier Point Road.

We had gotten off to a late start; the trail ran into a large patch of snow, and we had lost our way. Now it was pitch dark and we were stumbling ahead without a light, apparently along the left side of an invisible ridge.

The lion screamed again! Closer? I came to the sudden realization of fear. "We could die, Maryann. The lion might attack us, or we could walk off a cliff into oblivion." I was feeling the grave responsibility I held for leading her into this situation.

"We need to move ahead, but also move uphill," I said. "It'll reduce the possibility of falling." Maryann was one of the Flight Weenies who had worked on Minuteman III. Long blonde hair, blue eyes, and a slim figure. I hadn't talked with her much, but recently she had departed Autonetics and was now between jobs. At the going-away party, I had had the courage to ask for her phone number, and she had scribbled it for me on a scrap of paper. That was four months ago. I had finally met this wonderful woman, and we were joyfully sharing our lives.

We slowly made our way uphill. "I think it's getting not so steep now," she said. The taste of fear was still in my mouth—she was scared too. At last, we crested the ridge. "Look! A campfire!" We made our way down through the darkness on the other side.

"Hello," I said to the campers. "We're lost and heard a mountain lion. Is this Ostrander Lake?"

"Yes, it is. Not a great idea to be hiking at night." I asked if we could join them, and they said it was okay. Usually a loner, I felt relieved to be in the company of others. Tired and bone-chilling cold, we laid our sleeping bags down on our Ensolite pads, zipped them up together, and began to soak up each other's warmth. Happy to be alive, we promptly fell asleep.

* * *

Back at Autonetics, things were going poorly. The fifth Minuteman III launched from Vandenberg on October 15, 1969, FTM 305, and the tenth launched from Florida on December 10, FTM 210, had each failed to enter post-boost. I was greatly relieved to learn that MFS and the flight programs were blameless. It was later determined that particle short circuits—likely from the shock of third stage thrust termination, followed by weightlessness—allowed small metallic debris to float freely inside the PBCS electronics boxes where it wreaked havoc on the sensitive circuitry.

If I couldn't do physics, I could at least strengthen my knowledge about computers. Greg was my inspiration. We had wanted to drive to San Francisco the year before to attend the Fall Joint Computer Conference in hopes of learning something new. We had asked management for a few days of paid time off to attend the conference. Even though both Greg and I had hundreds of hours of unpaid "green time," our request was denied. This year, the conference would be in Las Vegas. I was looking forward to seeing the latest machines from the computer companies and having a good time eating the cheap food and walking through the gaudy casinos.

It was the night of November 18, 1969, on my way to Las Vegas, driving seventy-five miles an hour with the top down through the warm desert air— watching the waning gibbous moon out of the corner of my eye and listening to the radio. I heard *Contact light!* And a few moments later, *Man, oh man, Houston. I'll tell you, I think we're in a place that's a lot dustier than Neil's. It's a good thing we had a simulator because that was an IFR landing.* Apollo 12 had landed on the Ocean of Storms.

* * *

Suddenly, the decade was over. It was New Year's Eve, 1969. In the previous year, Larry and I had witnessed the first Minuteman III flight from Vandenberg. The Boeing 747 had taken flight—Larry and I drove to LAX to see the first of these giant planes pull up to the gate. The first Concorde supersonic airliner had taken flight, beaten by the Soviet TU-144 a few months earlier. Apollo 11 and 12 had landed on the moon. The Mariner spacecraft had passed by Mars, relaying detailed photos of the Red Planet. Not too far from me in California, the actress Sharon Tate and four others were slaughtered by Charles Manson and his "family" of women.

The bad news didn't stop. In February, Larry and his girlfriend Anne were in a terrible automobile accident, driving in Laguna Niguel, while looking for the new North American Rockwell building under construction; the new home of Autonetics. A car illegally crossed in front of them and Larry's red Datsun 1600 sports car slammed into it. Anne hit her head and was in a coma. I visited her in the hospital. Larry was all right, but he was beside himself with grief and worry. I was struck by the fragility of life. At any point, my tiny MGB could be hit by a cement truck, and I could die.

* * *

Trauma had clouded Greg's and Larry's lives, affecting me greatly. I felt lost, probably done with California. Meanwhile, the Flight Weenies continued to toil over what would become the final, definitive flight program.

Minuteman III hardware development had been conducted in a series of evolutionary blocks, with small improvements introduced between blocks. The first, Block I, had not been flight worthy, serving to test as much as possible while on the ground. We had been flying with Block II hardware, and now, in 1970, it would be Block III. Block IV was supposed to be the operational Minuteman III, to be replicated and deployed in a force of one thousand missiles. Up to now, the flight program had been custom-tweaked for each of the test flights. Soon, there would be only a single universal flight program serving for all the deployed missiles.

It was inevitable that the Block IV flight program would be subject to "formal software verification." Since MFS was the software tool used to verify the operation of the flight program, I was repeatedly asked by management if it actually worked. Recent launches closely tracked the MFS results. But did MFS always work with 100% accuracy all the time with every hypothetical variable? My answer was always, "I don't know."

The situation was complicated. Reference trajectories were supplied by TRW as a mathematical output from their own independent in-house physics simulation. MFS performed a physics simulation as well, but also simulated the instruction-by-instruction operation of the onboard D37D computer. The flight program flew the missile, attempting to follow the TRW reference trajectory, running inside of MFS. It seemed reasonable that no piece of software is perfect. Serious errors might be lurking in the TRW reference trajectory calculation, in the Minuteman Flight Simulator, or in the flight program itself—or even in all three parts of this software triad.

In the end, lacking any theoretical basis for determining the truth of the matter, a set of five trajectories/missions were selected by Autonetics, TRW, and the Air Force to be simulated using MFS to verify the flight program and give it the stamp of approval. The missions were specifically designed to exercise flight conditions, mission options, and trajectory shapes which "covered" the operational envelope. Various perturbations on such items as centers of mass, thrust profiles and winds were also introduced. An attempt was made to introduce these perturbations at points in the flights where conditions would tend to accentuate any adverse effects. Targeting data for these five simulated missions was developed by TRW, specifically to support the verification of the operational, Block IV, flight program.

Only the operational flight program, that is, the software used for the operational deployment of the weapon system, would be treated as a "hardware" end item on the Minuteman III system. In the end, the very expensive small roll of silver- and blue-colored Mylar tape encoding the operational flight program as a bunch of punched holes was what the Air Force bought for the taxpayers' money. Up to this point, the research and development flight programs had been treated as engineering tools and hence did not have formal demonstration, documentation, and sell-off requirements.

It was June, and finally there was a happy event for Greg's family. He and Marsha were getting married. I was deeply honored to be the Best Man at their wedding, renting a suit for the occasion. It was a typically bright sunny day with lots of flowers and smiles. I chatted with several notable guests, including the

232

father of the late actress Sharon Tate, who came in his late daughter's red sports car, and the North American Rockwell engineer who sat next to and coached Walter Cronkite on TV during the Apollo missions.

* * *

It had all been an incredible experience, but I had had enough. Studying physics at the University of Colorado had crept back in my mind. It was time to take the next step forward toward my Ph.D. For a year, I had been scrimping and saving to the bone. I had saved enough money to survive the expense of grad school—my frame looking pretty thin in the bathroom mirror.

Work on my computer had continued at a good pace. I was proud that two years after I had started working on it, my kitchen table held a computer that could perform two-hundred thousand additions per second. I had also purchased a surplus Friden Flexowriter with a tape reader and punch for input and output. I purchased a drum memory weighing two hundred pounds, salvaged from the bomber-era SAGE defense system, from C&H Surplus on Colorado Boulevard. C&H, in Pasadena, was the electro-mechanical equal to The Yard in Wichita. But I still didn't have a mobile robot to go with it all. That would have to come later.

I had foolishly hoped Maryann would come with me to Boulder. I dreamed about us taking camping trips with the MGB up in the Colorado mountains. But it was not to be. She was a California girl at heart and in the end didn't care enough for me to take such a leap of faith. I respected her decision and wished her well. I never saw her again.

I made plywood boxes for my computer, drum memory, Friden Flexowriter, drill press, and meager collection of tools, and had them shipped to my parents' house in Wichita. It was a bittersweet feeling turning in my badge and keys, leaving my friends and colleagues at Autonetics.

* * *

In July, Larry and I set out together on an epic Route 66 adventure, laying out our sleeping bags in national parks along the way to our final destination—Wichita. The relief from escaping the sterile confines of building 235 was tangible. We walked to the bottom of the Grand Canyon and back out, stayed at Bryce, and Carlsbad—we had a grand time.

As we laid out under the stars, we watched for meteors and talked for hours about wide-ranging subjects. Larry was studying at Cal State Fullerton to get an MS in business, plotting a future career that did not involve Minuteman. I asked him, "How's Anne feeling these days?"

"She's really doing all right, feeling much better. Thank God."

"Oh, that's good. I'll miss seeing her. I'm really going to miss Greg and his family. They were wonderful," I said. "Likewise, my friends Phil and Joan. You know, Joan fixed us great dinners on some memorable Sundays. And Rocky and Lee Ann. I don't think I ever told you Rocky saved my life. I was terribly screwed up after losing my wife and daughters. Rocky kept me from going insane."

"I'm sure I'm going to miss you, as well" he said.

I chuckled a bit. "I'm not so sure about that. Do you think anyone will ever find out about our sneaking in to watch the launch of FTM 301? Only four hundred yards from all that fire and smoke. You know, a few months later they flew me and some of the Flight Weenies up to Vandenberg in the Rockwell Aero Commander to see a launch from the blockhouse. It was not nearly as impressive as the launch we saw up close!

Another night, I said, "Do you remember my stories of amateur rocket building?"

"Sure," said Larry.

"I'm thinking it's freaking amazing how my middle school science project eventually led to the Minuteman Flight Simulator. Maybe I'll write a book about it someday."

Inevitably, we talked about women. At the time, Larry had two girlfriends. We tried to sing some verses of the popular song, *"Did you ever have to make up your mind?"*

"You know Larry, in the end, I've come up empty. All the women I've known have rejected me in one way or another. There was my wife, who left me. That put me into four years of certifiable depression. Do you remember Frankie, the English girl? No, that was way before we met. Anyway, she found a boyfriend and that was the end of our relationship. The next girl I spent a lot of time with was Georgie. I never had sex with either Georgie or Frankie."

"Why not?" asked Larry.

"I guess I was still hurting too much from my loss of Audine and my daughters. I don't think I ever told you about a girl who would show up at my door in the middle of the night. I forgot her name." I told a bit of that story to Larry who, amused, counseled me to avoid that kind of behavior. "I know, I know. She was only the second woman I had sex with in my life."

"I know you remember Becky, another of the keypunch operators. We were engaged for a while. But that didn't work out either. Of course, there was Maryann. I thought we had a good relationship."

"I remember you telling me how scared you were."

"Well, anyway, I still feel pretty good about myself. Each of the women taught me something—emotionally and physically. So in that regard, they each helped me a little bit to gradually get me out of my funk. All of the women were wonderful, and I owe each of them a debt of gratitude."

I said goodbye to Larry at the Wichita airport, wondering if we would ever be together again. I headed west once again in late July, this time driving the five-hundred-forty miles to Boulder where I did not know a soul. Before reaching Boulder, I stopped on the edge of a bluff and gazed down at a remarkable sight—the Flatirons looming like a bright vision over the city, the university, and my future.

* * *

The business of defending the country moved on. The first of a proposed one thousand Minuteman III missiles had been shipped to the Minot, North Dakota, Air Force Base missile complex on April 13, 1970, and was placed in a silo assigned to the 741st Strategic Missile Squadron—its flight program verified by MFS, the Minuteman Flight Simulator. It entered strategic alert status on August 19, 1970.

-oOo-

Epilog

At the University of Colorado

My twin goals were to graduate with a Ph.D. in physics, and hopefully find the love of a woman. The second came soon after arriving in Boulder when I met Beth Cawthra, The first required five years of effort with emotional support from Beth.

We were married in the Boulder County Courthouse on November 7, 1975. Eight days later in the physics department, I successfully defended my thesis, "Spin-flip Scattering from Free Holes in a Semiconductor." My bride, wearing her wedding dress, smiled proudly from the back of the audience.

WICHITA: Propulsion Research

The rocket boys of Propulsion Research got educated and headed out into the world pursuing interesting careers. **David Hollis** followed in our father's footsteps, becoming a successful architect, passing away from a long illness at age 72. **Harold Wiebe** received his master's degree in electrical engineering, after which he spent a year in Vietnam as a first lieutenant in the Army. Upon returning, he consulted with companies in the Cincinnati area, eventually becoming an associate professor of engineering technology at the University of Northern Kentucky. He passed away in 2020 at age 78. **Gary Peyton** became an expert in the chemistry of water, working with the University of Illinois in Champagne-Urbana. He passed away in 2018 at the age of 77. **Ron Gallop** loved flying. After three rides in the Tri-Pacer he was hooked on aviation, eventually becoming a pilot with Frontier Airlines and Ryan International Airlines Air Freight flying Boeing 727s and 737s. **Jon Freeman** graduated with a degree in aerospace engineering from the University of Colorado, pursuing a career in the design and operation of rocket engines, working with the mighty Saturn V Rocketdyne F-1 engines at Huntsville, Alabama. **Stephen Hawkins,** upon returning from his service in the Air Force, graduated in electrical engineering from the University of Wichita, pursuing engineering management at National Cash Register and Lear Jet Corporation in Wichita. **Phillip Roberts** graduated with a doctorate in physics from the California Institute of Technology. He joined the Jet Propulsion Laboratory in Pasadena, where he made seminal contributions to the design of gravity assist trajectories allowing spacecraft to visit multiple planets. Later, he contributed to laser guide stars that correct atmospheric distortion of light for telescopes. **Audine Taliaferro Hollis** remarried, had another daughter, and divorced again. She married once more and worked for many years as a lab technician. Our daughters grew up, now leading happily married lives. Carol became an attorney practicing corporate law with two sons. Dana became an occupational therapist, with two daughters and a son.

Minuteman III missile

Work continues on the Minuteman III missile to the present day. Immediately following the operational deployments of the weapon system in 1970, many improvements were made to guidance and control. Changes included adding the ability to program targeting remotely, introduction of a new PIGA leveling mode increasing robustness during seismic events and reducing the stable platform gyro drift rates by adding turbulence suppression hardware. Another improved platform alignment technique was developed, resulting in better gyrocompass accuracy, while eliminating all dependency on an external autocollimator. A change in the guidance algorithms increased the number of pre-stored targets, increased the off-azimuth launch angle from 14 degrees to 45 degrees, and improved gyro torqueing accuracy. Additional guidance improvements were initiated dealing with higher-order effects that had not previously been considered. After considerable efforts, it was seen these improvements had little effect on the circular error probable (CEP).

Initial deployment in Montana, North and South Dakota, Wyoming, Nebraska and Colorado reached 550 Minuteman III missiles by 1975, each with three RVs, along with 450 single-RV Minuteman II's, comprising a force of 1,000 missiles with 2,100 warheads representing approximately 820 megatons of destructive energy.

As early as 1969, Honeywell competed with Autonetics winning the contract to develop a replacement for the D37D which would incorporate a random-access memory and new associated electronics for interfacing with the

IMU. Production of this new flight computer did not begin until 2000, with final installation in 2007. The intricate Autonetics electromechanical IMU, the *heart* of the missile, remains with its six spinning masses.

Some of the **Mark 12 RVs** on the Minuteman III's were replaced beginning in 1979 with the **Mark 12A RV**, containing the W78 warhead, increasing robustness, safety, and the yield to 335 kilotons. The Mark 12A is heavier than the Mark 12, reducing the achievable range. For this reason, it was deployed in the northern-most missile fields at Minot AFB in North Dakota, and Malmstrom AFB in Montana. As a result of the **START** treaty, two RVs were removed from 150 of the Minuteman IIIs in 2001. The Obama Administration began de-MIRVing the remainder as part of the **New START** treaty. From June 16, 2014 onward, the Minuteman IIIs have only a single **Mark 21 RV**, containing the W87 warhead with a yield of 300 kilotons.

As of 2024, there are 400 single-warhead Minuteman III ICBMs deployed in Colorado, Nebraska, Wyoming, North Dakota and Montana. Another 50 silos can be loaded with stored missiles if necessary. All of the missiles are capable of restoration to three-warhead MIRV capability.

A new ICBM, the **MX Peacekeeper**, was produced from 1985 to 2005 with 50 deployed in Minuteman silos. The Peacekeeper could carry up to twelve **Mark 21 RVs** on its MIRV stage. The last Peacekeeper was retired in September, 2005, leaving Minuteman III as America's only operational ICBM with projected operation until 2030, after which an entirely new weapon currently called Sentinel, the Ground-Based Strategic Deterrent (GBSD), is proposed to take over.

Titan II missiles

Until 1987, 54 **Titan II** missiles remained deployed near Tucson, Arizona, Little Rock, Arkansas, and Wichita, Kansas, bringing a total force of 1,054 nuclear missiles. Each Titan II had a fourteen-foot-long **Mark 6 RV**, containing a W53 warhead with a yield of 9 megatons.

The Titans suffered three major accidents during their service. The first, and deadliest, was on August 9, 1965 at launch complex 373-4 near **Searcy, Arkansas**, when 53 workers in the silo were killed by toxic fumes. The second occurred when my father and brother Brian were working at the Hollis farm on August 24, 1978. They looked southward to see a red-orange cloud of oxidizer fumes a mile long, rising a thousand feet high, coming from launch complex 533-7, three miles away near the town of **Rock, Kansas**. Two crew members were killed when a poppet valve on the first stage oxidizer tank failed to seal during a fuel transfer operation, releasing a large quantity of toxic dinitrogen tetroxide. The third was on September 19, 1980, at launch complex 374-7 near **Damascus, Arkansas.** The socket from a worker's socket wrench rolled off a platform, hit a silo structure and ricocheted into the Titan's fuel tank, puncturing it. The resulting fuel leak eventually caused the missile to explode, destroying the silo, killing one man, and hurling the warhead into the sky, landing about 300 feet away. If the warhead had gone off, a large portion of Arkansas would have been destroyed. All the Titan II launch complexes were decommissioned and destroyed by 1987, except one near Tucson that serves as a museum.

Titan II was reborn as a space launch vehicle, serving the Gemini program and interplanetary missions into the 2020s.

Autonetics

In the mid-1960s, Autonetics had become the nation's largest military electronics complex, with an Anaheim campus of twenty buildings covering 188 acres and employing 36,000 workers at its peak. North American Rockwell became **Rockwell International** in 1973, renaming its aircraft division North American Aircraft Operations with Autonetics continuing in its role. In 1996, all of Rockwell International's defense and space divisions, including Autonetics, were sold to Boeing and integrated with Boeing's Defense division. Autonetics never occupied the million square foot Laguna Niguel facility, colloquially known as the "Ziggurat Building." Instead, it was sold to the federal government.

Now, most of the buildings of the Anaheim complex have been demolished and the property sold to developers. A single street name, Autonetics Way, survives. In 2010, Boeing erected a large monument at the edge of the site honoring the legacy of Autonetics, "Pioneering innovations in electronics and guidance, navigation and control that charted new frontiers on Earth and beyond." The monument is located at 3311 East La Palma Avenue. Lucille Randolph was a key contributor to its dedication.

In 2021, the Association of Air Force Missileers (AAFM) secured a video likely made in the early 1990s, entitled "Rockwell International—Making of a Winner—History of ICBM Guidance (Autonetics Division)." It was uploaded to YouTube and can be viewed at **https://www.youtube.com/watch?v=zIn-4sowLWI**.

Minuteman Flight Simulator

The question of whether MFS actually worked, remained. "Does it really accurately simulate every conceivable aspect of the missile and its environment, including every action that could be taken by the flight program?" I still have to answer, "I don't know." I was told a company based in San Pedro, California studied MFS for about a year. Then, another company, IBM Federal Systems Division in Los Angeles, worked on the problem. Finally, I was told that two graduate students at Brigham Young University in Provo, Utah studied the code, eventually concluding that it worked. I have no way of verifying any of this.

Kitchen table computer

Work on my kitchen table computer continued for several years. By 1975, the microcomputer integrated circuit was invented making it possible to carry a computer on a robot, powered by a battery. A mobile robot called Newt, the first with an onboard computer, was built by me and graduate student friend Dennis J. Toms while I was working on my Ph.D. The development was pictured on the cover of *Byte* magazine, highlighted in a 1978 WGBH Nova television program: *The Mind Machines*, and featured on the front page of the Wall Street Journal.

Cryolite and Hyper-Force rockets

My basement lab was disassembled to make space for Mom's sewing room. The MOUSE model, Firestreak rockets, and Arcturus were destroyed. Parts of the large rockets were stored at the farm in the barn's hay loft. These were discovered by the locals and carted off. Mother saved nearly all of the notebooks and drawings which were useful in writing this memoir.

Fallout shelter

Dad's proposed fallout shelter was never built. One issue may have been the cost of building it under the back porch. Or, it could have just been futile, given the rapid development of the enormous Soviet warheads.

WICHITA: Kansas personalities

Harold (Hal) Krier continued as an aerobatic champion, switching to a low wing monoplane based on the Canadian Chipmunk. His pioneering modifications of this aircraft were shared with other aerobatic pilots, notably Art Scholl. Harold was inducted into several aerobatic halls of fame. He died tragically in 1971 while spin testing a new airplane near Rose Hill, Kansas, a few miles from the Hollis farm. The airplane didn't recover from a spin, and his parachute failed to open. **Kansas Mack Sanders** eventually created a radio empire that included stations in nine markets, including Nashville. In 1978 he sold what was then Great American Broadcasting and began buying stations again. He died in 2003 and was posthumously inducted into the Country Radio Hall of Fame in 2005. **Tri-Pacer 8712C** continues to fly in the northern California area, having passed through multiple owners and converted to a tail dragger.

Retrospective

T HIS IS ME, the writer, no longer the fourteen to twenty-something youth of my story. It is 2024—with respect to my memoir, far into the distant future. The Cold War is long past, or so they tell us. But threats to the United States still abound. Minuteman III is still alive. Some four hundred nuclear armed Minuteman III missiles are still waiting in their silos for Armageddon; their gyrocompasses still spinning, still seeking Celestial North.

Looking back

Remembering the young person of my story, what do I see? I see a young boy who had the nerve to question the judgments of famous astronomers. I see someone who had a passion and the drive to realize his goals. I see a young student who did not do especially well in school but who was hungry to learn from books. He was a shy person, almost never participating in class discussions. I see a young man who substituted effort and hard work for intellectual achievement. A young husband and father who pushed ahead with his dreams that resulted in rejection and loss. I see a person who valued perseverance.

Maybe he was just ahead of his time. He built homemade rockets and a model satellite two years before Sputnik was launched. He could have waited five years to buy model rockets from Estes Industries. Nowadays, amateurs don't build rockets from scratch; everything needed to assemble powerful rockets is available online.

He used computers to simulate the rockets' flights long before they were widely available. He started building his own digital computer out of logic chips and a core memory. He could have waited nine years to buy Steve Wozniak's Apple 1 computer. But the young man could not wait. In writing this memoir, I found in the young man's 1967 notebook a handwritten quotation from Johan Wolfgang von Goethe: "Whatever you can do or dream you can, *begin* it. Boldness has genius, power, and magic in it." Many of his efforts, while ultimately not fruitful in themselves, were valuable learning experiences vital to his later career.

MFS and the *brain* of Minuteman III

One way to visualize the Minuteman weapon system is to picture it as a giant pyramid. At the base are the hundreds of launch facilities and the dozens of launch control facilities requiring perhaps ten thousand workers to construct and maintain. Next up on the pyramid are the rocket stages and re-entry vehicles, designed and built by five different companies, also requiring a large workforce to develop, manufacture, and maintain. Above that would be Autonetics' guidance and control system with its D37D onboard computer and IMU, carefully assembled by hand by perhaps a thousand people. At the top of the pyramid is the software that flies the missile and the software that checks the correctness of that software. Only a half-dozen people were needed to write the flight program, and less than a half-dozen to write the simulation, the Minuteman Flight Simulator (MFS), to verify the flight program's correctness. Notably, three of these people were women. We carved the tip of the spear, so to speak—the small roll of blue and silver punched tape.

Modern smart phones are at least ten thousand times faster than the onboard D37D flight computer, the *brain* of the missile. The same could be said for inexpensive single-board computers such as the Raspberry Pi.

MFS ran on IBM System 360 computers, requiring about 1.5 hours to simulate 9 minutes of powered flight. A modern smart phone is 50 to 100 times faster; in principle, it could run MFS in faster than real time. MFS was written mostly in the FORTRAN computer language (now written as Fortran), invented at IBM by John Backus in the mid-1950s. It is a general-purpose language for scientific and engineering work and remains in widespread use today.

Computer programming, for both the D37D and MFS, was incredibly crude compared with now. The lack of graphic displays and editing facilities forced programmers to carry much more of the design in their heads—engendering a vastly different thinking process from that of today's methods. This consideration, combined with non-interactive batch processing, led to a great deal of physical and mental fatigue.

The *heart* of Minuteman III

The intricate electromechanical *heart* of the missile, the Inertial Measuring Unit (IMU), especially the Pendulous Integrating Gyro Accelerometers (PIGAs), pushes the limits of physics in what is achievable. Today's cell phone solid state microelectromechanical IMUs, whose gyros use the Coriolis Effect, are no match for the Minuteman III's gyros that have a thousand times smaller drift rate.

The Draper 1977 annual report decries the trend that manual and mechanical skills are being replaced by computers that are supposedly infallible but are missing the innate talents of humans. When this happens, people skilled at making things are considered expendable. The report concludes that Draper continues to develop systems where craftsmanship is a principal ingredient.

MIRV and the ABM myth

Much has been written about the origins of Multiple Independently Targetable Re-entry Vehicles (MIRV). It seems the idea initially grew out of its technical feasibility developed in several places, including at Autonetics. It was also seen as an answer to a Soviet buildup of its anti-ballistic missile (ABM) capability which was thought might lead to a Soviet first strike that would wipe out the United States. MIRV gave Minuteman III an overwhelming strategic advantage over any feasible Soviet ABM system. As it turns out, the Soviet ABM was never a threat to the U.S.; MIRV effectively destabilized the balance of power, rapidly increasing our deployable nuclear arsenal with the possibility of achieving a U.S. first strike that could eliminate most of the Soviet Union's nuclear missiles, negating any significant retaliation. The Soviet Union did not deploy MIRVed missiles until 1975, five years after the first operational Minuteman III MIRV was deployed. Now, the Russian Federation, China, and France have nuclear missiles with MIRVs. Reportedly, India, Pakistan, Iran, and Israel have MIRVs under development.

Assured Destruction

"Assured Destruction" refers to the full-scale use of nuclear weapons by two or more adversaries which would cause the complete annihilation of both the attacker and defender. It was sometimes called "Mutual Assured Destruction (MAD)," but this was a misnomer since the Cold War policies of the Soviet Union and the U.S. were not symmetrical, and therefore were not "mutual."

Until the mid-1980s, the Soviets had a policy of "no first strike," that is, there was a pledge never to hit the U.S. with an overwhelming preemptive nuclear attack. It also embraced a policy of "no first use," that is, to refrain from using nuclear weapons in war unless first attacked by an adversary using nuclear weapons. Thus, the Soviet position was self-described as fundamentally defensive. The U.S. policy is also "no first strike," but it has never embraced "no first use."

Minuteman III, with its Post-Boost Vehicle (PBV) and the superb Autonetics guidance and control system has an extremely small Circular Error Probable (CEP). This led to early consideration of its use as a preemptive strike option to take out Soviet ICBMs, and certainly as a second-strike capability with pinpoint accuracy. Moreover, the accuracy of land-based Minuteman IIIs, with their stable launch conditions, exceeds that of the shorter-range Submarine Launch Ballistic Missiles (SLBMs).

As of June, 2020, the Russian Federation under Vladimir Putin released an amended policy outlining four scenarios that could warrant the use of nuclear weapons: (1) "in response to the use of nuclear weapons and other types of weapons of mass destruction against it" and/or (2) "in response to conventional attacks in situations critical to the national security of the Russian Federation." Two additional scenarios that did not appear in earlier documents include (3) an "arrival [of] reliable data on a launch of ballistic missiles attacking the territory of the Russian Federation and/or its allies" or (4) "attack by [an] adversary against critical governmental or military sites of the Russian Federation, disruption of which would undermine nuclear response actions." Thus, the Russian position can no longer be considered simply as defensive. China and the other nuclear powers severely complicate the nuclear picture.

In early 2022, policy discussions were underway to decide the fate of the Minuteman fleet. There are three options: (1) upgrade the system to sustain its effectiveness, (2) replace the fleet with a new ICBM referred to as the Ground-Based Strategic Deterrent (GBSD) in 2030, or (3) scrap the entire system. The initial plan favors option 2, now called the LGM-35 Sentinel. I believe the best choice is the first option, however, there should be a gradual reduction in the number of missiles, since there are already enough to destroy the world.

The New START treaty of 2011 with Russia was renewed by Presidents Biden and Putin in January 2021 — but only for a period of five years. In 2023, with Russia's invasion of Ukraine, Putin has threatened the use of nuclear weapons and has pulled out of New START. The treaty does not apply to tactical nuclear weapons. U.S. and NATO tactical forces possess several hundred B61 nuclear bombs that can be carried by F-15E and F16 fighters. Russia has a similar capability. The manned fighters can be diverted from carrying out a bombing mission, unlike the ICBMs and SLBMs which once launched cannot be recalled. One must assume that the use of a single nuclear weapon could rapidly escalate into all-out nuclear war.

For the U.S., the president has sole authority to authorize the use of nuclear weapons and does not need concurrence from military advisors or the U.S. Congress. Additionally, neither the military nor Congress can overrule these orders. If the president decides to order the launch of nuclear weapons, the nuclear "football" briefcase is opened, revealing war plan options developed under OPLAN 8010-12. These include major attack, selected attack, and limited attack options. The chosen option is combined with a Gold Code.

The Gold Code nuclear launch codes are arranged in a column, printed on a plastic card nicknamed the nuclear "biscuit." The card is similar to a credit card, and the president is supposed to carry it on their person. Before it can be read, an opaque plastic covering must be snapped in two and removed. The codes are changed daily. The list of codes on the card includes meaningless codes, and therefore the president must memorize where on the list the correct code is located. Selecting the correct code positively identifies himself or herself as the commander-in-chief and thereby authenticates the launch order. The selected Gold Code then propagates rapidly down the chain of command to the Minuteman III Launch Control Facilities (LCFs) and/or the SLBM submarines and nuclear bombers. The Secretary of Defense must jointly authenticate the order but cannot countermand it short of invoking provisions of the 25th amendment to remove the president from office.

It is assumed there are similar arrangements in the Russian Federation and Peoples Republic of China, and perhaps by the other nations that have nuclear weapons. This condition puts extraordinary pressure on the leaders of these countries to possess a sound mind; an extremely worrisome situation, since obviously it is not always the case.

A way forward

Physicist Freeman Dyson spent much of his life trying to find a solution to the problem of Assured Destruction. He offered suggestions in his 1984 book *Weapons and Hope*. After the tragedy of the two world wars, which began in quests for freedom, the development of large quantities of nuclear weapons obliviated any conceivable just cause. The result is that we are caught in a chain of tragic consequences.

Dyson proposes a doctrine he calls live-and-let-live: a determination to move away from nuclear weaponry in favor of advanced defensive and non-nuclear weapons. He believed this could be achieved by moral, political, and technical means, in that order. Morally, he compares resistance to nuclear weapons to resistance to the institution of slavery. Politically, relentless pressure must be brought to bear for international treaties. Technically, science and technology must push forward to develop effective non-nuclear weapons.

In many respects, it seems that this has been the path we have followed. We have reduced the number of strategic nuclear weapons and made important steps such as de-MIRVing the Minuteman III missiles.

According to the Federation of American Scientists 2022 report on the status of world nuclear forces, until recently the number of nuclear warheads has dramatically decreased from a peak of 30,000 Soviet warheads in 1961, and peak 40,000 U.S. warheads in 1991, to a few thousands of deployable warheads on each side. This progress is credited to the dissolution of the U.S.S.R. in the early 1990s as well as adherence to important arms reduction treaties between the two nations which control a combined 91% of the world's nuclear warheads. However, "All the nuclear weapon states continue to modernize their remaining nuclear forces, adding new types, increasing the role they serve, and appear committed to retaining nuclear weapons for the indefinite future."

Rogue states and terrorists

Nuclear weapons in the hands of terrorists are a distinct possibility. A cash-strapped nuclear-capable rogue state could sell a nuclear device to a terrorist group who could carry it on a ship and detonate it in a port or conceal it in a land shipment. Cyber-terrorism represents another threat, where a command-and-control center might be caused to inadvertently launch a nuclear missile.

In a 2020 *National Review* article, writer Brad Schaeffer contemplates the actions of self destruction which occurred on September 11, 2001. The attackers were so fanaticized that even bringing about Assured Destruction would not appear, to them, unreasonable. The power of the atom in the hands of suicidal fanatics could well be purchased on the black market by the highest bidder. He wonders if a technology that ended the cataclysm of the second world war will result one day in the end of life as we know it. If we are not hypervigilant, all our achievements as a civilization will be gone. It will be as if they never existed.

How I feel about what I did

I was a Cold Warrior working as a small cog in a giant human machine that produced an awesome weapon of mass destruction. To some people, I am guilty as charged; at least sometimes I have felt that way. My focus was like that of a surgeon doing the work, oblivious to the patient as a person. No one forced me to do it.

On the other hand, it has been 54 years now since the first Minuteman III was deployed and still no World War III. I feel pretty good about that. My work was for the greater good of the country. Building MFS was one of my most difficult achievements but also one of the thrills of my life. There is no question that I wanted it to succeed.

Source Materials

—Books & Booklets—

"Rockets and Jets," Herbert S. Zim, Harcourt, Brace and Company, New York (1945).

"Pocket Data for Rocket Engines," anon., Bell Aircraft Corporation (1955).

"Fallout Protection, What to Know and Do about Nuclear Attack," anon., Department of Defense Office of Civil Defense (December 1961)

"A Guide to Amateur Rocketry," anon., U. S. Army Artillery and Missile School, Fort Sill, Oklahoma.

"The Missile Next Door, the Minuteman in the American Heartland," Gretchen Heefner, Harvard University Press (2012).

"Ace in the Hole, the Story of the Minuteman Missile," Roy Neal, Doubleday & Company, 1961.

"Wichita's Legacy of Flight," The American Institute of Aeronautics and Astronautics—Wichita Section, with Jay M. Price (2003).

"Mayday Over Wichita, the Worst Military Aviation Disaster in Kansas History," D. W. Carter, the History Press, Charleston, S. C. (2013).

"Inventing Accuracy, a Historical Sociology of Nuclear Missile Guidance," Donald Mackenzie, MIT Press (1990).

"Making the MIRV: a Study of Defense Decision Making," Ted Greenwood, Ballinger Publishing Company (1975).

"Titan II, a History of a Cold War Missile Program," David K. Stumpf, University of Arkansas Press (2000).

"Weapons and Hope," Freeman Dyson, Harper & Row (1984).

—Documents—

"A Brief History of Minuteman Guidance and Control—March 1995," R. F. Nease and D.C. Hendrickson, March 17, 1995, revised May 3, 1995, Rockwell Defense Electronics. Autonetics internal publication, Anaheim, CA.

"Minuteman Flight Simulator," R. L. Hollis, Jr., L. W. Hambly, B. J. Kraus, and D. O. Dorius, Autonetics Technical Memorandum 70-242-005, April 3, 1970.

"Minuteman Weapon System History and Description," prepared for Intercontinental Ballistic Missile (ICBM) System Program Office (SPO), Hill AFB Utah, 84056, Ogden Air Logistics Center OO-ALC/LME by ICBM Prime Team, TRW Systems Prime – 19378, July 2001. [Obtained under the Freedom of Information act by Hans M. Kristensen].

"Software Validation Study," by D. Bruce Brosius. Autonetics AD-757 212, prepared for Space and Missile Systems organization, February 1973 [Distributed by National Technical Information Service, U. S. Department of Commerce, 5285 Port Royal Road, Springfield, Virginia 22151].

"Ballistic Missile Guidance and Technical Uncertainties of Countersilo Attacks," Matthew Bunn and Kosta Tsipis, Report Number 9, Program in Science and Technology for International Security, Department of Physics, Massachusetts Institute of Technology, 77 Massachusetts Ave., Rm. 20A-011, Cambridge, MA 02139, August 1983.

"Level II Documentation of Launch Complex 31/32, Cape Canaveral Air Force Station, Florida," Susan I. Enscore, Julie L. Webster, Angela M. Fike, and Martin J. Stupich, Historic American Engineering Record, December 2008.

"Inertial Navigation for Guided Missile Systems," Scott M. Bezick, Alan J. Pue, and Charles M. Patzelt, Johns Hopkins APL Technical Digest, Volume 28, number 4, 2010, pp. 331.

"The Evolution of Minuteman Guidance and Control," J. M. Wuerth, Navigation: *Journal of the Institute for Navigation,* Volume 23, No. 1, spring 1976.

"The Lure & Pitfalls of MIRVs, From the First to the Second Nuclear Age," Edited by Michael Krepon, Travis Wheeler, and Shane Mason, Stimson Center, May 2016.

"USAF Ballistic Missile Programs, 1967-1968," Bernard C. Nalty, Office of Air Force History, September 1969.

"Minuteman III Weapon System Description," T.O. 21M-LGM30G-I-13 (undated notes and diagrams).

"Case Study 3, The Origin of MIRV," typewritten notes, anon.

"Schuler Oscillations," Paul G. Savage, Strapdown Associates, Inc. SAI-WBN-14003, June 27, 2014.

"Minuteman III intercontinental ballistic missiles need not and should not be replaced because of the danger of their launch-on-warning posture," Frank von Hippel, APS Project on Nuclear Threat Reduction Advocacy Paper (November 14, 2020).

"Intercontinental Ballistic Missiles and their Role in Future Nuclear Forces," Dennis Evans and Jonathan Schwalbe, John Hopkins Applied Physics Laboratory National Security Report (2017).

"United States Nuclear Weapons, 2021," Hans M. Kristensen and Matt Korda, Bulletin of the Atomic scientists, 77:1, 43-63 (2021).

"Production Quality Assurance Testing of a Minuteman III LGM-30G Stage II Rocket Motor at Simulated Pressure Altitude (Motor S/N PQA6-60), W. D. Ervin and J. D. Gibson, ARO, Inc.

"Operating Manual for the Bendix G-15 General Purpose Digital Computer."

"Bendix Coding Manual for the Model G15D General Purpose Digital Computer, Bendix Computer Division of Bendix Aviation Corporation.

"Bits of Meaning, Introduction to the Bendix G-15," Bendix Computer Division, Los Angeles.

"Intercom Programming for the Bendix G-15 Computer," Linsley Wyant and John M. Howell, Wm. C. Brown Company, Publishers, 135 South Locust Street, Dubuque, Iowa, (1961).

"IBM 610 Auto-Point Computer," Form 23-6335-0 (8-57:5M-44951), International Business Machines Corporation, 590 Madison Avenue, New York 22, N. Y. (1957).

"Understanding Zinc-Sulfur Propellants," Antoon Vyverman, (2001 Edition).

"The Evolution of Earth Gravitational Models used in Astrodynamics," Jerome R. Vetter, Johns Hopkins APL Technical digest, Volume 15, Number 4, (1994).

"California dreamin'," Stuart J. Leslie, Physics Today **74**, 2, 36 (2021).

—Web Sites—

"The Minuteman III ICBM," https://nuclearweaponarchive.org/Usa/Weapons/Mmiii.html, accessed April 2021.

"50th Anniversary: First Minuteman III Placed on Alert at Minot," by Robert B. Cuthbertson, Jr., Minuteman III System Program Office, Air Force Nuclear Weapons Center / Published August 12, 2020, https://www.afnwc.af.mil/News/Article/2310952/50th-anniversary-first-minuteman-iii-placed-on-alert/, accessed April 2021.

"Aug. 16 marks 50th anniversary of first Minuteman III launch," by Bill Medema, Air Force Nuclear Weapons Center, Kirtland Air Force Base / Published August 15, 2018. https://www.kirtland.af.mil/News/Article-Display/Article/1603690/aug-16-marks-50th-anniversary-of-first-minuteman-iii-launch/, accessed 2021.

"Test Of Minuteman III ICBM With Three Reentry Vehicles Sure Seems Like A Warning To Russia (Updated)," by Joseph Trevithick, August 4, 2020, https://www.thedrive.com/the-war-zone/35352/test-of-minuteman-iii-icbm-with-three-reentry-vehicles-sure-seems-like-a-warning-to-russia., accessed 2021.

"Management and the Minuteman," Carl A. Jansen, The Space Congress® Proceedings. 1. (1965). https://commons.erau.edu/space-congress-proceedings/proceedings-1965-2nd/session-3/1.

"Minuteman Missile Nuclear Warheads," https://minutemanmissile.com/nuclearwarheads.html, accessed 2021.

"Russia Releases Nuclear Deterrence Policy," Shannon Bogos, Arms Control Association, July/August 2020, `https://www.armscontrol.org/act/2020-07/news/russia-releases-nuclear-deterrence-policy`, accessed 2021.

"Basic Principles of State Policy of the Russian Federation on Nuclear Deterrence," `https://www.mid.ru/en/foreign_policy/international_safety/disarmament/-/asset_publisher/rp0fiUBmANaH/content/id/4152094`, accessed 2021.

"Decision on Minuteman to shape US nuclear policy for decades," Associated Press News, `https://apnews.com/article/government-and-politics-415bef8de782dabc12520c9047ff5865`, accessed 2021.

"Status of World Nuclear Forces," Hans M. Kristensen and Matt Korda, Federation of American Scientists, March 2021, `https://fas.org/issues/nuclear-weapons/status-world-nuclear-forces/`, accessed 2021.

"The Future of the ICBM Force: Should the Least Valuable Leg of the Triad Be Replaced?" Ryan Snyder, Arms Control Association, March 2018, `https://armscontrol.org/policy-white-papers/2018-03/future-icbm-force-should-least-valuable-leg-triad-replaced`, accessed 2021.

Glossary

Accumulator	A special memory location for storing the results of a calculation
AFB	Air Force Base
Alignment blocks	Cube-shaped glass blocks with mirror surfaces (part of the IMU)
Attitude	Flight angle with respect to the air stream
Azimuth	Horizontal angle from North
Bendix G-15	Computer made by Bendix Corporation, introduced in 1956
Bearing	A machine element providing reduced friction for rotation
Bit	Binary digit, a 1 or 0
Byte	An 8-bit quantity representing 0 to 15
B-29	Boeing World War II four engine nuclear-capable bomber introduced in 1944
B-47	Boeing six engine nuclear-capable bomber introduced in 1951
B-52	Boeing eight engine nuclear-capable bomber introduced in 1955
C frame	Computational coordinate frame
CEP	Circular Error Probable, a measure of ICBM accuracy
CK-722	One of the earliest available transistors introduced in 1953

CO2	Carbon dioxide
CoM	Center of mass, origin of the MBF frame
de Laval nozzle	Convergent and divergent sections of combustion chamber through which exhaust gasses escape
DEC	Digital Equipment Corporation
Discrete	A single bit input or output to/from a computer
D-loop	A D37D computer special channel whose bits turn off or on the thrusters of the PBPS
D37D	Computer in Minuteman III PBCS
ECI frame	Earth-Centered Inertial coordinate frame
ECEF frame	Earth-Centered Earth-Fixed coordinate frame
Electrolytic level sensor	A 2-axis conductive bubble sensor for determining the direction of gravity while at rest (part of the IMU)
Ferrite	Magnetic ceramic used in core memories
Flight program	The program running in the D37D computer that flies the missile
FORTRAN	Now written "Fortran" is a general-purpose, compiled programming language that is especially suited to numeric computation and scientific computing originally developed by IBM in the 1950s for scientific and engineering applications.
FTM	Flight Test Missile
F-4C	Supersonic Phantom fighter-bomber produced by McDonnell-Douglas in 1960
F-100	Supersonic jet fighter made by North American Aviation in the mid-1950s
F-104	Starfighter produced in the late 1950s by Lockheed Aircraft Corp.
F-111	Fighter-bomber built by General Dynamics, introduced in 1967
G and C	Guidance and control
Gimbal	A rotatable ring-shaped structure with bearings permitting rotation around a single axis (part of the IMU)
G frame	Gyro coordinate frame
Gyro	A rapidly spinning mass that resists disturbing external forces (part of the IMU)
Gyrocompass	A gyroscope that determines True North (part of the IMU)

HICS	Hardened Intersite Cable System
H-loop	A D37D computer special channel which outputs information to control the Minuteman III thrusters
IBM 026	IBM keypunch that prints on the card top edge
IBM 360	Groundbreaking IBM computer system first delivered in 1965
IBM 610	IBM computer introduced in 1957
IBM 650	A more advanced computer than the 610, introduced in 1954
IBM 729	IBM magnetic tape system
IMU	Inertial Measuring Unit
IBM 1410	IBM decimal computer introduced in 1960
IBM 1620	A transistorized IBM computer with a ferrite core memory, introduced in 1959
IBM 7094	A transistorized IBM computer with a ferrite core memory, introduced in 1962
ICBM	Intercontinental Ballistic Missile
IGY	International Geophysical Year
KC-135	Boeing four engine jet tanker introduced in 1957
LAX	Los Angeles International Airport
LF	Launch Facility: Underground "silo" housing a single Minuteman missile
LCF	Launch Control Facility: Underground manned control center for a connected field of ten LFs
LCEF frame	Launch-Centered Earth-Fixed coordinate frame
Lbs.	Pounds (Imperial units)
LITVC	Liquid Injectant Thrust Vector Control
Mach number	Ratio of a vehicle's speed to the local speed of sound
Matrix	A block of number having rows and columns
MBF frame	Missile Body Fixed coordinate frame
MIRV	Multiple Independently-Targetable Re-entry Vehicle
Megaton	Unit of explosive power equivalent to weight of TNT
MFS	Minuteman Flight Simulator
MOL	Manned Orbiting Laboratory
MOUSE	Minimum Orbital Unmanned Satellite of the Earth
MS frame	Missile Station coordinate frame
NDEA	National Defense Education Act (1958)
Octal	A 3-bit quantity encoding numbers from 0 to 7
PA-22	Four-seat Tri-Pacer airplane made by Piper Aircraft Company (1950-64)

PDP-8	Early minicomputer produced by the Digital Equipment Corporation
PBCS	Post-Boost Control System
PBPS	Post-Boost Propulsion System
PBV	Post-Boost Vehicle (PBPS+PBCS)
PIGA	Pendulous Integrating Gyro Accelerometer (part of the IMU)
PIGA frame	PIGA coordinate frame
P frame	Platform coordinate frame
Pitch	Rotation around the "belly" of the missile
PRIDE	Personal Responsibility in Daily Effort
PTRC	Harold Wiebe's Parachute Timing Release Computer (1957 design)
Register	A special memory location for storing important quantities in a computer
Resolver	A device for measuring very small changes in rotation (part of the IMU)
Runge-Kutta	The integration method used in MFS
R-loop	A D37D computer special channel which inputs angle information from the gimbal resolvers
Roll	Rotation around the central axis of the missile
ROTC	Reserve Officer Training Corps
RV	Re-entry vehicle
R7	The Soviet Union's rocket that launched Sputnik
SAT	Self Alignment Technique
Semi-somnus	The condition where the D37D computer halts computations during an X-ray burst
Stable platform	The gimbal-mounted physical structure of the IMU upon which the two platform gyros, three PIGAs, two level sensors, and two alignment blocks are mounted
Stage	One of MMIII's solid propellant or liquid propellant sections
STL	Space Technology Laboratories
SC-4020	Stromberg-Carlson pioneering printer-recorder
SINS	Ships Inertial Navigation System
Subroutine	A program invoked by a higher level program
Theodolite	Instrument for measuring elevations at several different azimuth angles
TITAN II	Large liquid propellant ICBM produced by the Glenn L. Martin company, in service from 1962 to 1987

Torquer	Electromagnetic device for applying torque to gimbals (part of the IMU)
TRW	Thompson Ramo Wooldrige, Inc.
T-34	Low wing trainer aircraft introduced by Beech Aircraft Company in 1953
U-loop	A D37D computer special channel which computes the precise fine-countdown time for thrust termination or object deployment
Vernier engine	Small auxiliary rocket engine used for adjusting the direction of thrust
VIDEO	Betty Jo's Validation and Demonstration program
V-loop	A D37D computer special channel which inputs velocity information from the PIGAs
VOR	Very high frequency omnidirectional range system
Word	Basic computational unit, consisting of a group of bits representing a number
XB-70	Six engine Mach 3 nuclear bomber, built by North American Aviation in 1964
Yaw	Sideways rotation of the missile

www.ingramcontent.com/pod-product-compliance
Lightning Source LLC
Chambersburg PA
CBHW060907120626
46553CB00001B/239